Vavilov and his Institute

A history of the world collection of plant genetic resources in Russia

by **Igor G. Loskutov** (Vavilov-Frankel Fellow - 1993)

The International Plant Genetic Resources Institute (IPGRI) is an autonomous international scientific organization, supported by the Consultative Group on International Agricultural Research (CGIAR). IPGRI's mandate is to advance the conservation and use of plant genetic resources for the benefit of present and future generations. IPGRI's headquarters is based in Rome, Italy, with offices in another 15 countries worldwide. It operates through three programmes: (1) the Plant Genetic Resources Programme, (2) the CGIAR Genetic Resources Support Programme, and (3) the International Network for the Improvement of Banana and Plantain (INIBAP).

The international status of IPGRI is conferred under an Establishment Agreement which, by January 1998, had been signed and ratified by the Governments of Algeria, Australia, Belgium, Benin, Bolivia, Brazil, Burkina Faso, Cameroon, Chile, China, Congo, Costa Rica, Côte d'Ivoire, Cyprus, Czech Republic, Denmark, Ecuador, Egypt, Greece, Guinea, Hungary, India, Indonesia, Iran, Israel, Italy, Jordan, Kenya, Malaysia, Mauritania, Morocco, Norway, Pakistan, Panama, Peru, Poland, Portugal, Romania, Russia, Senegal, Slovakia, Sudan, Switzerland, Syria, Tunisia, Turkey, Uganda and Ukraine.

Financial support for the Research Agenda of IPGRI is provided by the Governments of Australia, Austria, Belgium, Brazil, Bulgaria, Canada, China, Croatia, Cyprus, Czech Republic, Denmark, Estonia, F.R. Yugoslavia (Serbia and Montenegro), Finland, France, Germany, Greece, Hungary, Iceland, India, Ireland, Israel, Italy, Japan, Republic of Korea, Latvia, Lithuania, Luxembourg, Macedonia (F.Y.R.), Malta, Mexico, Monaco, the Netherlands, Norway, Pakistan, the Philippines, Poland, Portugal, Romania, Slovakia, Slovenia, South Africa, Spain, Sweden, Switzerland, Thailand, Turkey, the UK, the USA and by the Asian Development Bank, Common Fund for Commodities, Technical Centre for Agricultural and Rural Cooperation (CTA), European Union, Food and Agriculture Organization of the United Nations (FAO), International Development Research Centre (IDRC), International Fund for Agricultural Development (IFAD), International Association for the promotion of cooperation with scientists from the New Independent States of the former Soviet Union (INTAS), Interamerican Development Bank, United Nations Development Programme (UNDP), United Nations Environment Programme (UNEP) and the World Bank.

Citation:
Loskutov, Igor G. 1999. Vavilov and his institute. A history of the world collection of plant genetic resources in Russia. International Plant Genetic Resources Institute, Rome, Italy.

ISBN 92-9043-412-0

IPGRI
Via delle Sette Chiese 142
00145 Rome
Italy

Contents

Foreword

The Vavilov-Frankel Fellowships Programme was established by the IPGRI Board of Trustees to commemorate the unique contributions to plant science by Academician Nikolai Vavilov and Sir Otto Frankel. The first two Fellowships were awarded in 1993 to Igor Loskutov from Russia and Robin Pistorius from the Netherlands. The research of the 1993 Fellows focused on examining and recording the work of scientists in the plant genetic resources community, including those after whom the Fellowships Fund is named.

This research has produced two books, the present one by Igor Loskutov and a complementary one, published in 1997, by Robin Pistorius: Scientists, Plants and Politics – A History of the Plant Genetic Resources Movement. Together, these two volumes aim to describe and analyze the historical background to today's efforts to conserve and use plant genetic resources. They are a tribute to the early pioneers in this field and a record of the actions and debates that have done so much to shape the way the scientific community addresses conservation today, as well as the perspectives of the many other individuals around the world who are equally concerned with the maintenance of this priceless genetic heritage.

Igor Loskutov undertook meticulous research into records including the unique archives of the Vavilov Institute and interviewed key scientific figures including Sir Otto Frankel who knew Nikolai Vavilov and visited him in 1935. As the present book was being finalized in late 1998, we learned sadly of the death of Sir Otto. We are grateful for his contribution to both this book and to that of Robin Pistorius, and salute his readiness to share memories, ideas and opinions with younger scientists.

Geoffrey Hawtin, Director General, IPGRI

Acknowledgements

This publication became possible through the author's being awarded an IPGRI Vavilov Frankel Fellowship named after the prominent researchers of this age in the sphere of plant genetic resources. This book is the first attempt to give a comprehensive overview of historic events of the past 100 years, regarding the efforts of the N.I. Vavilov Institute of Plant Industry (VIR) in plant genetic resources collecting and its international activities in this area. It is based on numerous materials: articles, memoirs, Vavilov's works, his extensive correspondence and other publications (mainly in Russian) edited for the most part in recent years in commemoration of N.I. Vavilov's centennial and the 100th anniversary of the Vavilov Institute of Plant Industry.

Work within this project in close cooperation with Robin Pistorius from the University of Amsterdam enabled us to present a different, "westernized" angle to various scientific and political problems associated with the Institute's activities. I am extremely grateful to him for his useful counsel and priceless assistance as well as for his hospitality during our meetings in Amsterdam and for the aid he provided during our joint trips to Rome (IPGRI) and Canberra (CSIRO). I am also pleased to express my sincere gratitude to Lyndsey Withers for the excellent organization of all the work within this project and to other colleagues in IPGRI: E. Frison, D.H. van Sloten, J. Engels, J.A. Dearing and M. Nocca.

It was a pleasure and honour to meet the late Sir Otto Frankel in Canberra (CSIRO) where he shared his memories about his meetings with Nikolai Vavilov and gave me useful counsel, which I appreciate. I am deeply grateful to Professor J.G. Hawkes whom I met in St. Petersburg for his permission to use his memories of his meeting with Vavilov in this book. I am grateful to the responsible officers of the genebanks in the former COMECON countries: S. Goral and W. Podyma (Poland); R. Koeva and I. Alekseev (Bulgaria); L. Holly and L. Horvat as well as A.T. Szabo (Hungary); L. Dotlacil, I. Bares, Z. Stehno and A. Kovachik (Czech Republic); K. Hammer, H. Knupffer, P. Hanelt, T. Gladis and K. Pistrick (Germany); G. Erdenejav (Mongolia).

I would like to express my gratitude and compliments to Dr Yury N. Vavilov, N.I. Vavilov's son, for his assistance, for his advice and comments, for photographs and letters from his private archive and for the documents he found in the National Archives of USA (Washington, DC), which I used in the present publication. I also greatly appreciate the valuable help of Ms R. Harvey, Archivist of John Innes Centre (UK) which I received in my work with the archives of W. Bateson, A.D. Hall and C.D. Darlington; Ms T.K. Lassan, Archivist of VIR for her assistance in my work with documents and photos of Vavailov's times, the workers of the Science and Technology Archives of St. Petersburg (Russia); and Fay Warner for assistance with editing the manuscript.

I express my deep gratitude to Prof. V.A. Dragavtsev, Director of VIR, for his confidence in me when he nominated me as a participant in this project and for his assistance in my work. I am grateful indeed to Mr S.M. Alexanian, Head of the Foreign Relations Department (VIR), for he helped me a lot in organizing my trips during the project and made available materials related to the international cooperation of VIR. I also appreciate his comments on the manuscript. Special thanks should be addressed here to A. Kh. Bakhteev for his aid in selecting illustrations and rare publications for this work. I wish to extend my thanks to the leading researchers of VIR who helped me in my efforts, specifically to Prof. S.N. Bakhareva, Prof. V.D. Kobylyansky, Dr O.P. Mitrofanova as well as to the translators of VIR – Mr A.G. Krylov and Mr S.V. Shuvalov.

Igor G. Loskutov, St. Petersburg

1 Collecting and studying plant genetic resources in Russia from 1894 to 1920

Establishment and activities of the Bureau of Applied Botany before 1905

The intensive development of agriculture in the 1870s and 1880s promoted an increased interest in agronomy and agricultural science in general in the Russian Empire. In the sphere of plant production, this interest pursued purely practical objectives related to describing, preserving and utilizing populations and races of local cultivars.

When studying the flora and plant life, botanists fixed their attention almost exclusively on wild species, while the majority of cultivated plants were not studied. The works of Koernike, Seringe, Alefeld and others laid the basis of the taxonomy of cultivated plants. These works were important for use in agriculture and practical plant breeding. Professor A.F. Batalin, who was the first to study cultivated plants in Russia, encountered materials of immense richness and repeatedly stressed the necessity of setting up a special laboratory for applied botany that would concentrate on the comprehensive study of the Russian cultivated flora. This idea was supported by other researchers, for instance, by Professor A.S. Famintsin and Professor I.P. Borodin.

To carry out these tasks a Bureau of Applied Botany was organized in 1894 under the Scientific Committee of the Ministry of Land Cultivation and State Property. In compliance with its Regulations, the Bureau comprised three departments: Information, Scientific and Acclimatization. These departments were responsible for determining the taxonomy of both cultivated and wild plant species. They supplied information on the sources where those species were available, and studied both cultivated and wild plant species regarding their botany, agronomy and phytopathology. They also promoted the introduction of various plant species and varieties into cultivation (Regel 1915).

Professor Alexander F. Batalin was the first director of the Bureau of Applied Botany and its only staff member He was an outstanding botanist and crop taxonomist, and the founder of applied botany in Russia. At the same time, he acted as director of the Imperial Botanical Gardens in St. Petersburg. However, when he was director of the Seed Testing Station of the St. Petersburg Imperial Botanical Gardens, Batalin on his own initiative had already started collecting and researching varietal diversity of rye, spelt, millet, rice, buckwheat, legumes, common onion, oil-bearing Cruciferae and other crops. This research formed a basis for further activities of the Bureau.

Because of the vast amount of work, he was unable to pay due attention to the Bureau's activities. Additional funds were allocated to maintain the Bureau's work, which was often fragmentary. In the first years of its existence, the Bureau of Applied Botany acted as an information service responding to requests concerning the crops cultivated in Russia. It could hardly be regarded as a scientific institution.

The Bureau mostly concentrated on reviewing various books, proposals and projects which were submitted to the Scientific Committee of the Ministry of Land Cultivation.

After Batalin's death in 1899 Professor Ivan P. Borodin, a Russian botanist and a member of the Scientific Committee, became director of the Bureau of Applied Botany. At the same time, he was also professor at the Institute of Forestry, a full member of the Academy of Sciences, director of the Botanical Museum, and editor of the Proceedings of the Naturalists' Society. Thus, the many responsibilities of Borodin, in combination with a

Fig. 1. A.F. Batalin, the first Director of the Bureau of Applied Botany [VIR Archives].

Fig. 2. I.P. Borodin, Director of the Bureau of Applied Botany [VIR Archives].

shortage of funding, as well as absence of premises and equipment inhibited the development of the Bureau's activities in this period.

In order to improve the Bureau's work and to stimulate the Scientific and Acclimatization Departments, at the end of 1900 Borodin invited Robert E. Regel to become a member of the Scientific Committee. Before 1900 Regel had many contacts concerning training with western botanists, such as Engler and Asherson at the Higher School of Horticulture in Potsdam. Regel read lectures on the application of botany to horticulture in St. Petersburg University and he was a researcher at the St. Petersburg Imperial Botanical Gardens (Flyaksberger 1922).

From 1901, practically all the work of the Bureau lay upon Regel's shoulders. Despite very scanty funds allotted to the Bureau at that time and a lot of routine work that distracted Regel from his major responsibilities, he nevertheless managed to collect and study Russian barleys. To acquire seeds, between 1901 and 1904, letters were dispatched to all Russian provinces. These letters contained a request to collect and forward seeds and ears of local barley varieties. As a result, in these years over 990 barley samples were collected from all regions of Russia. The most interesting samples were received from the Northern Caucasus and Transcaucasus. The richest diversity was found in areas adjacent to modern Iran and from Armenia. In addition to Russian barleys, the first foreign accessions were obtained at that time. For example, at the Bureau's request V. Sounders from the Central Experiment Station in Ottawa sent a collection, while the "Atterberg's collection", rich in diverse barley forms, was acquired from Sweden. All accessions were described taxonomically according to Koernike, and studied for their morphology and the nature of the grain (Regel 1915).

Fig. 3. E.R. Regel, Director of the Bureau of Applied Botany (1911) [VIR Archives].

Fig. 4. N.I. Litvinov, researcher of the Bureau of Applied Botany responsible for oats (1911) [VIR Archives].

Fig. 5. A.I. Maltsev, researcher of the Bureau of Applied Botany responsible for weedy plants (1911) [VIR Archives].

The structure and organization of the Bureau in the years 1905-1920

In 1905 Regel became director of the Bureau of Applied Botany. From the very beginning of his joining the Bureau, Regel had been trying to combine the descriptive knowledge of a botanist with practical horticulture and agronomic objectives of the Bureau while also studying cultivated plants. Developing the principles laid down by Batalin and Borodin, Regel introduced his complex approach toward the study of both cultivated and weedy plants. He attempted to apply taxonomic principles to cultivated plants, having linked these principles with the agronomic characters of the crops. That was the feature of his approach that distinguished it from those of foreign researchers. In this respect, Regel can be called the founder of scientifically based applied botany, or applied taxonomy. Setting forth his principle that "applied botany is a special botany of cultivated and field crops and weedy plants", Regel targeted his activities at the development of this section of botanical science. He proceeded from the assumption that:

> ... not only botanists, but also agronomists, foresters and farmers – whatever agricultural sphere they may be involved in – must be distinctly aware of what plants they are studying, observing or cultivating; otherwise their researches, observations and agricultural practices will inevitably lack a sound foundation (Flyaksberger 1922, p.5).

Regel tried to set up the Bureau's work on a purely scientific basis. He deemed that the Bureau should not be preoccupied with popularization and teaching activities, as it would impede the pure scientific research. Regel himself studied barley. Unlike other plant taxonomists, he recognized two species of cultivated barley, *Hordeum vulgare* and *H. distichum*. He identified over 54 new constant lines of barley isolated from local populations, which were tested by field planting. Regel discovered barley forms with smooth awns and described them in a monograph. He also started studying the problem of protein content in Russian barley seed and improved the adaptability of six-row winter barley for brewing, so that it could replace the previously used two-row barley from Europe (Flyaksberger 1922).

In 1906, when the Bureau's barley collection and the results of its study were presented at the International Exhibition in Milan, the highest prize (Diploma d'Onore) was awarded to the Bureau of Applied Botany. The results were summarized by Regel in *Les orges cultivées de l'Empire Russe*. In 1907 there were certain changes in the management of the Ministry of Agriculture. Prince Boris B. Golitzyn, a very energetic and distinguished statesman, was appointed Chairman of the Scientific Committee. Under his leadership and with the direct involvement of the members of the Scientific Committee the plan of reorganization was developed for various bureaux under the control of the Committee The accent was on the practical objectives of their scientific activities which were to receive funding from the state budget. In this respect, the Bureau of Applied Botany in addition to plant genetic resources studies became responsible for the systematic collecting of cultivated, weedy and wild plants in Russia (Regel 1917a).

From this time the statute of the Bureau of Applied Botany was amended by excluding the issues associated with fungus diseases of plants, since an independent Bureau of Mycology and Phytopathology was founded under the directorship of a well-known plant pathologist Professor Artur A. Yachevsky (Regel 1915).

After 1907 the Bureau's financial status improved to some extent; this enabled Regel to resign from his additional offices and continue his plantings not only at his own estate in the Caucasus, but also in Kursk Province. A qualified agronomist was invited to supervise these plantings. Research responsibilities concerning other crops of applied importance such as wheat, oat, pasture grasses, weeds, sunflower, millet, etc. were distributed by Regel among new employees. That autumn he gained an opportunity to employ Konstantin A. Flyaksberger for permanent service as his assistant, to whom a study on Russian wheats was entrusted. Next year Nikolai I. Litvinov and Alexander I. Maltzev were appointed on a permanent basis. Litvinov begun a study on Russian oats, and Maltzev on the field weeds of Russia. In that period the Bureau begun to study pasture grasses (Regel 1915).

By 1908 the Bureau had at its disposal specialists with high qualification and it became possible for the Bureau to pass from making requests for plant samples to the full-scale implementation of its major objective: the systematic exploration, collecting and detailed studying of cultivated and weedy plants of Russia. During the period of 1908-1909, the St. Petersburg, Kursk, Liflandian[1], Don and Poltava regions were explored. Collection of plant materials and surveys of these areas was performed with the participation of R. Regel, as Head of the Bureau, A. Maltsev, as an expert in weeds, and K. Flyaksberger, as an expert in wheat diversity (Shcherbakov *et al.* 1971).

Among foreign guests of the Bureau, Professor N.E. Hansen from the USDA should be mentioned. In 1908 he paid his third visit to the institution in order to familiarize himself with the recent barley introductions. His first visit was devoted to collecting alfalfa in Turkestan, the second one to collecting fruit crops in Siberia, and in 1908 he paid a visit to the Bureau on his way to Persia where he planned to collect forage grasses and alfalfa. In previous years at his request he had been supplied with new barley varieties described by the Bureau (Maltzev 1909).

For the purpose of plant collecting, other experts from USDA visited Russia: M. Carlton (1898-1899) to collect wheat and other major cereal samples; S. Knapp and M. Carlton (1900) to collect local varieties of wheat, forage plants and other agricultural crops; and E. Bessy (1902-1914) to collect local alfalfa populations and fruit-bearing species noted for winter hardiness. F. Meyer, a famous plant introduction expert, attached great importance to collecting activities in Russia as a source of local plant diversity.

Throughout the 10-year period of 1905-1915 Vavilov explored European and Asiatic (Siberia) regions of the country. In 1909 Regel undertook a tour of European countries. He visited the Berlin Botanical Gardens, the Higher School of Horticulture at Dahlem and the Biological Institute of Agriculture and Forestry in Germany, the Copenhagen Botanical Gardens in Denmark, and while in Sweden, he became acquainted with the Stockholm Botanical Gardens and the Svalof experiment station. Regel's journey allowed the Bureau to strengthen contacts with foreign botanical and breeding institutions as regards exchanges of seed and published materials and to present the Bureau's activities to the fullest extent. That very year, the Bureau of Applied Botany fulfilled several requests for seeds submitted by the Svalof experiment station, the Stockholm Botanical Gardens, the German Seed Growing Association, and experiment stations in Switzerland and Holland (Maltzev 1910).

After 1910 the Bureau began to train research fellows in the field of applied botany, barley, wheats, oats and weeds. Among these trainees was a young specialist Nikolai Vavilov. When Flyaksberger came to the Bureau, Regel's dream of publishing a *Bulletin of the Bureau for Applied Botany* became a reality. The Bulletin focused on the activities of the Bureau, and also contained translations of foreign books and articles. All articles were published with extended summaries in a foreign language. The Supplements to the Bulletin contained monographs by employees of the Bureau. Among the foreign authors were G. Schull, E. Baur, H.C. Fruwirth, G. Mendel, H. Sieben, A. Diels and others. Besides

---

[1] Now comprising Lithnania, Latvia and Estonia.

various research materials, the Bulletin published many papers on the methodology of cultivated plant studies. In many respects this increased the popularity of the Bureau in Russia, as well as abroad. The Bulletin was the first in the world to deal with applied botany and cultivated plants. Since 1908 when publication began, its issues have been mailed to breeding and botanical institutions within Russia, as well as to Holland, the USA, Canada, Germany and Sweden. Among private contributors to the Bulletin there were prominent botanists, plant breeders and geneticists of that time such as G. Nilsson-Ehle and A. Atterberg from Sweden; E. Baur, F.A. Koernicke, A. Zade and A. Schulz from Germany; H.C. Fruwirth from Austria; J. Percival from Britain; A. Thellung from Switzerland; W. Saunders from Canada; L. Trabut from Algeria, etc. (Regel 1915).

In acknowledgement of the Bureau's activities by the scientific community, the 2nd Russian Congress on Breeding and Seed Growing (1912) claimed the *Bulletin of the Bureau for Applied Botany* to be the central organ to be published in the sphere of applied botany and plant breeding in Russia, and recommended that this publication be renamed *Works on Applied Botany and Plant Breeding*.

Regel had inspired the use of scientifically appropriate initial material for the breeding of plants. This was reflected in the activities of the Bureau, in publications of the Bulletin and in the reports on the Congress of Plant Breeders.

It should be noted that in 1908 N.I. Litvinov published *Guidelines for Production of Uniform Cereal Plantings during Comparative Botanical Research*. In this work the author presented a summary of the research of the Bureau and gave the first methodological guidelines for research on botanical diversity of cultivated plants, specially modified for the purposes of applied botany. From 1911, the collections were assembled primarily by means of the direct collecting of seed and plant samples by Bureau staff. Examination of the collected material was conducted at both the Bureau's experimental fields and at the collecting sites. Within the period 1911-1914, collecting of seeds and herbarium samples was carried out in the previously explored St. Petersburg, Liflandian and Kursk Regions, as well as in Moscow, Novgorod, Samara, Voronezh, Perm, Taurida, Kharkov and Kherson Regions. In this period Regel, Maltzev and Flyaksberger were joined by V. Kuznetsov, F. Satsyperov and other researchers of the Bureau as participants in collecting missions (Shcherbakov *et al.* 1971).

By 1912 the Bureau employed seven leading researchers supervising the major crops in Russia, and oil-bearing crops, including sunflower. The expansion of the Bureau's scientific activities required a widening of its provincial experimental network. The Voronezh Branch (in South Russia) in the steppe zone and the Novgorod Branch in the northern forest area, 200 km from St. Petersburg, became the main experimental sites of the Bureau. The Bureau carried out planting on private lands rented for a number of years. In order to identify and test the inheritance of different crop characters, these experimental plots were chosen in regions of extremely different climatic and geographical conditions (Regel 1915).

Objectives and tasks of the Bureau, 1905-1920

By its 20th anniversary in 1914 the Bureau of Applied Botany had become a famous and respected institution in the sphere of crop diversity studies both in Russia and abroad and had its own specific objectives and methods. The Bureau's basic task was to study cultivated crops and useful, weedy or detrimental wild plants of the Russian Empire. Special research projects were conducted on the following agricultural crops: all cereals (wheat, barley, oat, rye, millet, *Panicum Sorghum*, rice, etc.); industrial crops including fibre and oil-bearing plants and others; horticultural crops (cabbage, cucurbits and melons, legumes, root crops, tuber crops, medicinal and aromatic plants, and fruit-bearing plants); as well as wild plants, such as all weeds and pasture plants (grasses, sedges and legumes). This research was performed by applying scientifically precise methods including detailed studies of the scientific literature for the subject under investigation.

By 1914 the Bureau's collections had been greatly enlarged by accumulating the germplasm requested and shipped from various farms in Russia and by the collecting missions of the Bureau's researchers. In 1914 the major collection contained: wheat, 4100 accessions; barley, over 2900; oat, over 1000; rye, about 400; pasture grasses, about 500; sunflower, more than 450; weeds, over 1000; the carpological collection, more than 1000 seed samples, and other plants, over 2000 accessions. All in all, there were 14 000 accessions. The Bureau's herbarium contained more than 10 000 specimens collected in different provinces of Russia.

At this time, the Bureau's basic priorities were to determine the composition of local varieties, their taxonomic names and geographic areas of distribution. The most important practical result of these activities was collecting, identification and description of the diversity of plant varieties cultivated in the Russian Empire, which helped to restore the diversity of cereal varieties and populations (in particular barley) that had been lost during devastating droughts in the Volga basin. All the above-mentioned crops were described in terms of the composition of local populations with respect to the races, forms and species present. The Bureau's study of most of these crops resulted in identifying the inherited morphological and agronomic characters, performing crosses and understanding the genetical basis of diversity of some of the crops along Mendelian lines. Complex examination of these collections led to the development of taxonomic systems in several major crops, generated from the data acquired by studying the heritability of morphological, anatomical, cytological, biochemical, immunity and agronomical characters (Regel 1915).

When World War I started the Bureau suffered a decline in its activities. Many Bureau workers joined the Army, the funding was reduced both in the Bureau's headquarters in Petrograd and in its provincial branches, but the Bureau continued its scientific research. Different aspects of various crops were studied at both the Bureau's branches and the central facility. The Bureau continued publishing the results of its research on different collections. For example, a paper by A. Maltzev entitled *On the development of wild and weedy oats* appeared in 1914, K.A. Flyaksberger's *Identification of wheats* and *Wheat varieties in Siberia* (Flyaksberger 1915a, 1915b) and N. Litvinov's *On* Puccinia glumarum *incidence in spring wheats in Kamennaya Steppe in 1914* in the following year (Litvinov 1915), A. Maltzev's *Weeds in the governate of Novgorod* (Maltzev 1916b) and F. Satziperov's *The hybridization of* Helianthus annus *L. x* H. agrophyllus *A. Gray.* in 1916, and R. Regel's *On species formation* and Vavilov 's *On the origin of the cultivated rye* in 1917 (Maltzev 1916b).

In 1916 the Bureau of Applied Botany was renamed the Department of Applied Botany and Plant Breeding. At the end of 1917 Vavilov was recommended by R. Regel, K.A. Flyaksberger and A. Maltzev to be promoted to the position of Assistant Head of the Bureau. Vavilov attributed major importance to the activities of the Bureau and especially to Regel, Head and actual reorganizer of the institution. In one of his letters of 1924, Vavilov wrote:

> Applied botany as a separate field of science that deals with cultivated plants has been introduced into Russia by Robert Regel. Before him, this trend was developed by Batalin and Koernuke, but the latter has left Russia and transferred his research to Germany. The greatest merit of Regel is that having started the work all alone in his study, he developed it to such extent, that at present applied botany has become widely practiced

> and is considered a scientific field of paramount importance at all experiment and breeding institutions in Russia.
>
> R. Regel not only devoted all his life to this science, but also involved quite a scientific force and many practical workers. The Bureau of Applied Botany established by him has been transformed into the Department of Applied Botany and Breeding at the State Institute of Experimental Agronomy, which is known throughout the world. The journal Works on Applied Botany and Breeding founded by Regel is the only one that deals with this field of agricultural knowledge.
>
> Regel was the permanent head of the Department until his sudden death in 1920 of typhoid caught while on a business trip to Moscow (Vavilov 1980, p.157).

In 1920, following Regel's death, N.I. Vavilov was elected Head of the Bureau of Applied Botany. Thus, 1920 marks the end of the first stage in the formation and development of the Department as an institution involved in collecting, studying and conserving plant resources.

2 Brief sketch of Vavilov's life up to 1920

Education in Moscow

Nikolai Ivanovich Vavilov was born in Moscow on 25 November (13 November in the pre-Revolution Russian calendar) 1887. Vavilov was the child of a prominent textile manufacturing family. His father, Ivan Il'ich Vavilov, of peasant background, became a wealthy textile merchant and leader in the Moscow City Council. Despite wealth, the family lived simply, emphasizing the virtues of modesty, hard work and self-discipline. From his parents he acquired the habit of wearing unpretentious clothes and eating simple food. With his brother, Sergei, a physicist and later President of Academy of Science of the USSR, his sisters, Alexandra a physician and Lidia, a microbiologist, he obtained his early education in Moscow.

Following his father's wishes, Nikolai Vavilov went to the Commercial School of Moscow and graduated in 1906. Then he enrolled in the Moscow Institute of Agriculture (Sinskaya 1991).

In 1917 Vavilov wrote in his *curriculum vitae*:

> In the spring of 1911, I graduated from the Moscow (Petrovsky) Agricultural Institute (now the Timiryazev Agricultural Academy in Moscow) and I remained at the Department of Farming under Professor D.N. Pryanishnikov to prepare a professorship and worked as assistant at Moscow Plant Breeding Station in 1911 (under Professor D.L. Rudzinsky). After graduating I worked some months in 1911-1912 at the Bureau of Applied Botany in St. Petersburg (under Professor R.E. Regel) and at the Bureau of Phytopathology and Mycology (under Professor A.A. Yachevsky) (Vavilov 1980, p.29).

During his work at the Moscow Agricultural Institute, Vavilov became interested in plant breeding and the work of Gregor Mendel. In 1910 he worked on the Poltava Experimental Station (Ukraine) set up to provide peasants with better small grain varieties and to conduct systematic studies of disease resistance problems, concerning oats, wheats and barley. In the same year, Vavilov started working as a trainee at the Breeding Station of the Moscow Agricultural Institute and became involved in experimental research on plant immunity.

In October 1911, in his letter to Regel, Vavilov wrote concerning his request to become acquainted with scientific work of the Bureau as a unique institution combining research on the taxonomy and geography of cultivated plants (Vavilov 1980). From 1912, he was charged with convening practical seminars at the Golytsin Higher Agricultural Courses for Women. It was there that he delivered a lecture entitled *Genetics and its Linkage with Agronomy* where he posed a new problem of interrelation between scientific theory and practice. He spoke about the origin of a new science of genetics, about Mendel's works, the mutation theory of Korzhinsky and de Vries, Johannsen's principle of pure lines, and wide possibilities in obtaining new crop cultivars on the basis of Mendel's theory (Vavilov 1990).

Fig. 6. Researchers of the Bureau of Applied Botany (1911). In the front row: N.I. Vavilov (trainee), A.I. Maltsev, E.R. Regel, K.A. Flyaksberger and N.I. Litvinov (private archive of A.Kh. Bakhteev).

Fig. 7. Researchers of the Moscow Breeding Station (1912). In the centre: N.I. Vavilov, sitting; behind him D.I. Rudzinsky and to the right S.I. Zhegalov, both standing [VIR Archives].

Studies abroad

In order to be prepared for a research career, Vavilov was sent abroad for two years in 1913. For a prompt establishment of contacts with foreign scientists, he was supplied by R.E. Regel with reference letters. The tasks undertaken by Vavilov for his two-year trip abroad included acquainting himself with the activities of leading experts in plant breeding and genetics in Great Britain, Germany and Austria. He also intended to attend lectures at Cambridge University and in Germany and planned to dedicate half of his second year abroad to America where he wished to study the organization of applied botanical research, visit biological institutes and get acquainted with the works of the leading experts in those fields. Unfortunately, a greater part of this programme failed owing to the outbreak of World War I.

In 1913 he was offered a period abroad, which he spent mainly in Great Britain. He was trained by William Bateson at the John Innes Horticultural Institute at Merton near London and under Sir Rowland Biffen at the School of Agriculture in Cambridge. He also used the personal library of Darwin, which was located in the Botany School of the University of Cambridge (Vavilov 1980).

He was deeply impressed by W. Bateson, both as a scientist and as a democratic administrator who allowed his staff great freedom to pursue their individual scientific ideas and experiments. Vavilov described in detail his impressions of Bateson's laboratory in Merton in his article *W. Bateson. 1861-1926* (Vavilov 1926b). In particular, he wrote:

> Contrary to a traditional concept of the reticent English personality, it was impossible to imagine a more cordial attitude, attention and readiness to help, encountered by a young Russian researcher in Merton (Vavilov 1926, p.511).

He also spent much of his time with V.H. Blackman at the Institute of Plant Physiology in London and with R. Punnett and R. Biffen at Cambrige University advancing his study on the immunity of cereal plants to diseases. He also visited the geneticist Haeckel in Germany and was told about the plant breeding company Vilmorin in France during this period. When World War I broke out he returned to Russia, after his 14 months abroad (from August 1913 until October 1914) (Vavilov 1980).

Activities between 1914 and 1920

Upon his return Vavilov committed himself with an even greater zeal to his research on disease resistance, genetics and breeding at the Plant Breeding Station of the Moscow Agricultural Institute. Vavilov was exempted from army service because of an eyesight deficiency. He was able to continue his scientific and teaching activities, unceasingly expanding the scope of his research on cereals, wheat in particular, and recruiting younger scientists, trainees and students (Bakhteev 1988).

When Vavilov was studying the wheat collection at the Breeding Station of Moscow Agricultural Institute, he identified a sample of wheat with extremely high resistance to powdery mildew. This sample known as 'Persian wheat' was a form of bread wheat. After experimenting on crossing this sample and wheat species of different ploidy, Vavilov found out that the chromosome number of this sample was 28, while the bread wheat had 42. In 1928 Vavilov isolated this sample into a separate species and named it *Triticum persicum* (Vavilov 1980).

His study of crop resistance to fungal diseases led him to feel the need for fundamental investigations into the taxonomy of cultivated plants, and the latter resulted in approaching the problems of experimental genetics. In order to enrich the collections of cereal crops at the Breeding Station, in 1915-1916 Vavilov undertook brief trips to the Trans-Caspian region and Turkmenistan, adjacent to Persia (Iran), during which he kept on collecting and thoroughly studying local wheat varieties with the hope of finding more samples of the resistant wheat species so that its origin could be traced (Vavilov 1987b).

Supposedly, the routes of the first of Vavilov's trips had been prompted by the results of works by Bateson who undertook a trip to Russia in 1886 and explored the territories of present-day Turkmenistan and Kazakhstan to collect the diversity of cultivated plants.

First collecting expeditions

In 1916 Vavilov carried out his first large-scale cultivated plant collecting mission to Asia and explored northern Iran, adjacent Russian territories and the Pamir Mountains. In those years (as he wrote in *Five Continents*) World War I was still on, and the Russian army advancing towards Turkey, occupied a considerable territory of northeastern Iran. The troops located in the Astrabad, Mazenderan and Gilyan Provinces quite often suffered a kind of disease similar to alcohol intoxication.

In order to determine the reasons for such a disaster, the Russian Ministry of Agriculture followed the advice of the Moscow Agricultural Institute and sent Vavilov from Southern Turkestan to Iran. Having examined wheat varieties in northern Iran, Vavilov found them to be badly contaminated with *Lolium temulentum* L., a highly toxic weed, and susceptible to fusarium wilt. Hot bread (eaten immediately after baking) made from this wheat caused symptoms similar to drunkenness. The reasons for the soldiers' sickness became clear enough, and special orders were issued forbidding the use of local grain.

Taking advantage of this commission, Vavilov commenced exploring the cultivated flora of northern Iran. He failed to find Persian wheat there. Upon completing his trip in northern Iran, Vavilov proceeded to the central part of the country, pursuing the same objective. The volume of collected wheats, barleys, ryes and other crops kept on growing with every day. His remarkable findings considerably enriched the knowledge of bread wheat taxonomy. For the first time, the amazing association of wheat variation with the ancient centres of diversity became obvious to Vavilov. The fields of winter wheat in inner Iran turned out to be strongly contaminated with primitive rye. Frequently, especially with increased altitude, rye was replacing wheat. For the first time Vavilov put forward a hypothesis of cultivated rye development from its weedy state which had contaminated the original ancient wheat. This concept was reported by Vavilov in 1916 to the Russian Botanical Society and further published in the *Bulletin of the Bureau for Applied Botany*.

Having completed his trip to Iran in August 1916, Vavilov decided to keep on exploring the cultivated plants during the remaining part of the season. He travelled along the border between Russia and Afghanistan and proceeded to the centre of the Pamirs. Here he found completely unknown endemic forms of wheat, rye, peas, vetchlings and French lentil, most of which proved to be powdery mildew resistant (Vavilov 1986a).

Vavilov's journey to Iran and the Pamirs determined to a great extent the planning of his further explorations. The role of mountainous regions of southwestern Asia became evident. The abundance in those regions of a great number of wild crop relatives, i.e. wild barley, *Aegilops* species, wild lentil, wild rye, showed that the most fascinating and the most complex mystery of evolution might be unveiled. Vavilov became more convinced of the necessity of penetrating still deeper into southwestern Asia, into Afghanistan, Chitral, Nuristan (Kafiristan), and into southwestern India.

After a successful collecting mission Vavilov returned to Moscow and in spring 1917 he launched an impressive programme of field maturity experiments with the collected materials.

Saratov period

In 1917 Vavilov was appointed as Professor in the Department of Agriculture (in the narrow sense) and Plant Breeding at the Saratov Higher Agricultural Classes which in 1918 was reorganized into The Agronomic Department of Saratov Agricultural Institute. In the autumn of 1917 Vavilov moved to Saratov. At the same time, he was elected Assistant Head of the Bureau of Applied Botany but postponed taking office until 1 September 1918 (Vavilov 1980).

Vavilov brought with him to Saratov rich and diverse experimental plant materials, over 6000 samples in all, accumulated through the years of teaching the Golitsyn Agricultural Classes, working at the breeding station of the Moscow Agricultural Institute, and collecting in Iran, Turkestan and Pamir. Young Professor Vavilov was accompanied to Saratov by the graduates from the Moscow Agricultural Institute, namely O.V. Yakushkina, A.Yu. Freyman (Tupikova), E.A. Stoletova and E.N. Sinskaya who assisted him in his field experiments (Bakhteev 1988).

Vavilov's lectures were very popular with the students, as the scope covered was far beyond the subject of agriculture (in the narrow sense) and included plant breeding, genetics, modern plant production, history of agriculture, and utilization of plant resources in plant introduction and breeding practices. In his introductory lecture, Vavilov told the students:

> The study of the geography of cultivated varieties in Russia has just begun... The wild flora has been little studied as a source of various plants that may be domesticated...(Vavilov 1965, p. 438).
>
> However rich nature may be in forms, the combinations of characters that would perfectly suit man would be extremely rare, and the deliberate creation of new and agriculturally more advanced forms constitutes a current objective of plant science... The recent experiments in genetics have unveiled much more opportunities than a researcher of the past could only dream about... In the near future man will be able to synthesise forms completely unimaginable in nature...(Vavilov 1965, p. 440).

Thus, the immediate objectives were formulated.

In his field experiments Vavilov always had scores of voluntary assistants from amongst his students who were fascinated by the magnitude of his ideas. In spring 1918, at Regel's initiative, Vavilov founded the Saratov Branch of the Bureau of Applied Botany where he launched field experiments on the premises of the Saratov Agricultural School (transformed into the Saratov Agricultural Institute at that time). That spring Vavilov planted over 12 000 plots with various hybrid wheats and barleys, as well as the materials gathered on collecting missions. These plantings formed the basis for the graduation papers of his students. Noteworthy enough, these plantings were organized and made by Vavilov immediately after the October Revolution of 1917, i.e. in the spring and autumn of 1918.

In the prevailing conditions of post-war destruction and depression Vavilov's work and his optimism were to be of enormous importance. In a letter to R.E. Regel in 1918 Vavilov wrote:

> The plot at the farm of the Agricultural Institute is safer because of its distance from the soldiers' camp. So far, the troops are not abundant, but we expect an increase in the near future, and therefore the sowings are endangered. Last year the sunflower was completely destroyed at the station (Vavilov 1980, p. 34).

In 1919 and 1920 Vavilov was forced to transfer all the plantings from the Saratov Breeding Station to a farm 8 km from Saratov. In addition to the lines and samples acquired and collected by Vavilov, the samples from the collections of the Bureau were multiplied at this site, which determined the large scale of sowings in the following years.

It should be noted that in 1917-1920 the group of student researchers headed by Vavilov made Saratov practically the only scientific centre in Russia dealing with plant science, breeding, and genetics. The report of the Agricultural Scientific Committee for the period of 1918-1920 stated that:

> ...of the local establishments, only the Saratov Branch for Applied Botany at the Saratov University carried out works properly and in full volume. During the three recent years the activities have even been broadened by including a series of new crop plants..., and by expanding genetic research on cultivated plants (Vavilov 1987a, p. 399).

At that time the study of smooth-awned barleys launched in 1915 at the Breeding Station of Moscow Agricultural Institute was going on. Vavilov paid a lot of attention to various taxonomic aspects of wheat. Having studied the rich collections at the Moscow Breeding Station, Bureau of Applied Botany, and later the collections of Percival in Britain and Vilmorin in France, Vavilov became convinced that bread wheats were represented by a great diversity of forms. He continued the study of wheats collected in Iran, Turkestan and Pamir in 1916 and finally identified many new forms. Because of this, Vavilov understood the poor development of bread wheat taxonomy. The results of this work were reflected in a paper *On the understanding of bread wheats: (Taxonomic and geographical essay)* published in the *Works on Applied Botany and Plant Breeding*.

The research on cereal disease resistance was continued in Saratov as well. A great number of samples collected by Vavilov himself and received from other sources were tested against natural and artificial backgrounds.

In 1918, while in Saratov, Vavilov summarized his works on plant protection problems in his publication *Immunity of Plants to Infectious Diseases*. Dr R.E. Regel, reviewing the process of Vavilov's work, wrote:

> In the past 20 years the problems of immunity have been studied by many distinguished scientists in almost all countries of the world, but I dare say that nobody has ever approached the solution of these complex problems with such a breadth of views and comprehensive coverage of this problem, as Vavilov has done. His voluminous work on immunity, which he has been gradually preparing for publication, will evolve into an undoubtedly remarkable work, an honour to Russian Science in the whole world's scientific community (Vavilov 1987a, p. 419-420).

Disease resistance studies started by Vavilov when he was a student led him to a thorough taxonomic research on genera, species and intraspecific taxa of cultivated plants and their close wild relatives. He noticed that despite a striking diversity of forms, their variability is governed by strict regularities. Undoubtedly, Vavilov started developing the concept of parallel variation in species long before he came to Saratov. But, only there did he manage to complete this work, thanks obviously to the availability of voluminous breeding and introduced material, and to the continued help of his laboratory assistants.

Thus, he discovered and formulated the *Law of Homologous Series in Variation* on which he gave a report at the Third All-Russia Plant Breeding Congress. This law states that closely related species, genera and families reveal parallel variation of all the traits characteristic of them. It reflects the similarity of hereditary variation in neighbouring taxonomic units and helps to foresee biological traits and agricultural characters from analogy with related species and genera (see also Vavilov 1920).

Vavilov's report at the Congress was highly valued by the participants as an outstanding scientific event, and as a tremendous achievement by the young scientist. The participants appreciated the Congress as a historical one and declared Vavilov to be "The Mendeleyev of biology". The law of homologous series had a similar importance for applied botany and plant breeding as did the system of elements for chemistry. The first results of Vavilov's fundamental research alone proved Regel's testimonial of Vavilov in 1917 as a talented scientist who would become the pride of national science.

Between 1917 and 1921, when he was head of the Saratov Branch of the Department of Applied Botany and Plant Breeding, Vavilov organized some local expeditions on the territories of the Volga River. In August 1920, with a group of students and teachers from the Agriculture Department of Saratov University, Vavilov undertook an explorative voyage down the Volga River to Astrakhan with the aim of collecting and studying local varieties of cultivated plants. Further on, they explored the delta of the Volga, and on their way back, they again explored neighbouring areas and made a tour of Elton and Baskunchak lakes (Korotova 1978).

The results of a comparatively brief exploration of the cultivated flora of southeastern Russia together with other materials accumulated in over three years at Saratov enabled Vavilov to publish *Field Crops of the South-east* in 1922 (Vavilov 1922c). In this paper he reviewed the field crops of the southeast from the point of view of a botanist, and presented the results of three years of observing a multitude of crop varieties collected in various countries, as well as of different crops grown in Saratov, Samara, Astrakhan and Tsaritsyn Regions.

Theoretical concepts of Vavilov's research – new advances from De Candolle and Darwin

The idea of researching the problems of the origin of cultivated plants arose in Vavilov's mind when he was still a student. While working at the Moscow Agricultural Institute, in addition to his lecturing Vavilov spent much time working with an extended group of students interested in natural science. Even in the second year of his work with the Institute, he gave a lecture at one of the group's sessions on the topic *Genealogy of the Plant Kingdom of Plants*. In the summer of the same year he took part in a students' expedition to the Caucasus. In 1909, at the gala meeting of the Institute dedicated to Charles Darwin's Centennial, he presented a report entitled: *Darwinism and Experimental Morphology*. In 1912, he published his above-mentioned lecture as *Genetics and Its Linkage with Agronomy* (Bakhteev 1988).

The views of young Vavilov had been greatly influenced by the activities of the Bureau of Applied Botany. For the first time Vavilov met with Regel, director of the Bureau in 1911 in Kharkov (Ukraine) at the First Congress on Agricultural Plant Breeding and Seed Growing. Later that year, Vavilov wrote in his letter to R.E. Regel:

> At the Kharkov Breeding Congress you let me hope for assistance; now dare I repeat again my humble request to let me work with the Bureau and assist me in getting acquainted with its activities... As a task for myself, I would like to set a more or less detailed acquaintance with the work of the Bureau, so far the only institution in Russia that combines studies in taxonomy with those on the geography of cultivated plants; most of the time I'd like to dedicate to cereal grass the taxonomy, i.e. familiarizing myself with the major literature sources, identifying difficulties in the determination of cultivated cereal grasses, and surveying the Bureau's collections. I would regard any advice from a bureau staff member regarding the use of your library as really valuable to me (Vavilov 1980, p. 18).

In 1917, in another letter to Regel, Vavilov wrote:

> Applied Botany attracted me even when I was a student. Although I have been taught most of the time in Russia and abroad by plant pathologists and geneticists, I can identify myself as an applied botanist and have as it seems the strongest bonds with the applied botanical community (Vavilov 1980, p. 30).

During his brief practice at the Bureau of Applied Botany young Vavilov listened with great attention to the comments of R.E. Regel, K.A. Flyaksberger, A.A. Yachevsky and to a few other researchers of the Bureau.

In the course of his years with the Breeding Station of the Moscow Agricultural Institute, under the prominent Russian plant breeder D.L. Rudzinsky's supervision, Vavilov started his studies on cultivated plants. It was then that he understood Darwin's theory of evolution and profoundly analyzed A. De Candolle's *L'origine des plantes cultivées*.

In a letter of 1923 Vavilov wrote:

> After reading De Candolle, Darwin and Gehn a dozen times, I feel quite deservedly that we have managed to get a bit ahead of De Candolle (Vavilov 1980, p. 136).

The development of his ideas of the origins of cultivated plants was most clearly set out in his article *The Theory of Origin of Cultivated Plants After Darwin*. Vavilov wrote that when in 1913-1914 he studied in Darwin's private library he had an opportunity to see what persistence Darwin revealed in studying the works of his predecessors in the history of plant domestication and breeding. In the same publication he remarked that Darwin's approach to the variability and evolution of cultivated plants was, first of all, based on De Candolle's *Géographie Botanique Raisonée*. However, unlike the latter, Darwin was mainly interested in the evolution of species and hereditary variations of a species under cultivation. De Candolle's basic concern was in identifying the origin of cultivated plants. Vavilov considered that:

> The great and immortal merit of Darwin is the fact that he drew attention to the dialectics of interrelated variation, heredity and selection, thereby opening up opportunities for progressive breeding (Vavilov 1965, p. 159).

After Darwin's death, the book *L'origine des plantes cultivées* by De Candolle was published and became the basic work in this sphere. De Candolle's classical work filled with an enormous quantity of facts seemed to Vavilov one-sided, i.e. revealing only the aspect of the primary origin of cultivated plants and their links with the original or related wild species. Following Darwin, not De Candolle, Vavilov paid much attention not only to the major areas of species' origin, but also to the evolutionary stages passed by plant species in their distribution under the effect of the domestication processes, environments, and natural and artificial selection. Proceeding from the basic tenets of Darwin's and De Candolle's theories, Vavilov formulated the long-term research objectives which constituted the foundation for the activities of his Institute. They were described by him in the same article:

> Proceeding from Darwin's geographical and evolutionary theories, systematic studies of the most important plants were planned, including all the evolutionary stages from the initial areas, where the linkage with wild types could be traced and where the phylogenetic relation between different wild species and cultivated forms can be established, via the

> subsequent further historical dispersal of the species up to the final links with contemporary plant breeding. This work could, of course, be done only by a team of scientists working according to a definite plan based on the principle of evolution (Vavilov 1965, p. 162).

Consequently, the investigations covered the major part of the agricultural areas all over the world, though paying attention mainly to the Soviet Union itself and the adjacent countries. All the agricultural areas of North and South America underwent thorough comparative study. The Cordilleras in South America, the major agricultural mountain range of Africa and a considerable portion of the agricultural areas in the Asian continent were also studied. Enormous collections of material were made, amounting at present to over 200 000 viable samples, which became subject to all-round study, and were sown or planted under various conditions with the application of all of the modern methods available.

> Not only did these studies double our knowledge of various cultivated plants, but also led to the discovery of a large number of new species and an enormous amount of botanical varieties of both cultivated crops and their close wild relatives, all of which had previously been unknown to botanists and, even less so, to plant breeders (Vavilov 1965, p. 163).

Vavilov ended this work with the following words:

> Darwin's theory of evolution is a cornerstone, being the basic and unique theory that has been standing solidly for more than 80 years. In their professional activities, botanists, zoologists, geneticists, plant breeders and ecologists as well as plant geographers are influenced by this universal theory, and it is only due to this fact that understanding the process of evolution and the functioning of organisms becomes possible (Vavilov 1965, p. 176).

3 Vavilov as Director of the Institute of Plant Industry

Vavilov's initiatives in the Department of Applied Botany and Plant Breeding

In 1920 Vavilov was offered the position of Head of the Department of Applied Botany and Plant Breeding. Only in March 1921 did he manage to move to Petrograd together with about 20 young researchers who had been assisting him in Saratov. Prior to his arrival in Petrograd[2], in a letter to G.S. Zaitsev late in 1920 he expressed his vision of the reorganization of the department:

> Plans are numerous. I would like the Department to be a necessary institution, as useful to everybody as possible. I'd like to gather the varietal diversity from all over the world, bring it to order, turn the Department into the treasury of all crops and other floras, and launch the publishing of "Flora Culta", the botanical and geographical study of all cultivated plants. The outcome is uncertain, especially considering the surrounding hunger and cold. But still, I want to try (Vavilov 1980, p. 79).

After the revolution of 1917, in the ensuing circumstances the move from Saratov, with its orderly lifestyle and natural farming at the Station, to Petrograd was an heroic act. The food situation in Petrograd was dramatic with food rationing, to say nothing of the heating which was a problem for everyone, for the houses were mostly heated by small stoves known as "bourgeoikas". Before moving out from Saratov, Vavilov wrote to Berg:

> Such a move from the south-east would seem unnatural to many, but this is not so obvious, and "man doth not live by bread alone... (Vavilov 1980, p. 41).

Encountering many insoluble problems, after his move, Vavilov wrote:

> Myself, I have been in Petrograd for a week. There are millions of troubles. We are fighting against the cold at home, and for furniture, flats and food.... I must confess that it is quite a problem to arrange a new laboratory and an experimental station as well as to settle 60 employees. I am accumulating patience and persistence.
>
> About three weeks will pass before we settle everything, and then the planting season will approach. We shall need horses, tools, workers... If only we manage to set things going here. But setting the work going will be much more difficult here than in Saratov. The life here is anyway much harder, especially today (Vavilov 1980, p. 41).

Travels in America and Europe

Though busy with providing proper conditions for his researchers in Petrograd (at 44, Hertzen Street) and setting up an experimental station at Tsarskoye Selo (later named Detskoye Selo; now Pushkin, near Petrograd), Vavilov was preparing to attend the International Phytopathological Congress in New York which was held in August 1921.

---

[2] St. Petersburg was renamed Petrograd in 1914; Petrograd became Leningrad in 1924, reverting to the original name of St. Peterburg in 1991.

During this trip Vavilov had a practical task, as prior to his departure he had been appointed as an expert for the food and crop seed purchasing commission. Vavilov wanted to restore the old contacts in Europe and establish new ones in the New World, as well as getting acquainted with the current state of research abroad, and acquiring scientific literature. The voyage across the Atlantic lasted almost a fortnight. Though physically tortured (he could not stand sea voyages as he suffered badly from seasickness) Vavilov found strength to revise and extend an English version of the *Law of Homologous Series in Variation* for British publishers. This work was reported by him at the International Congress. It made a very strong impression on the participants and was widely promoted in the press, which was a help to Vavilov in later days (Vavilov 1922a).

After this, Vavilov started studying the scientific achievements of the USA in biology and applied botany. His first visit was to the University of Columbia, where Thomas Morgan, a world-famous geneticist, was working. In Washington, DC Vavilov visited the USDA Bureau of Plant Industry. Here he studied the routes explored and the plants collected by such "plant hunters" as Fairchild, Meyer, Hansen, Harlan and others. After his visit to the Washington Bureau, Vavilov came to the conclusion that the results of botanical collecting missions could have been more substantial if a scientifically based theory of introduction of domesticated plants and their wild relatives had existed. Later, in his *Five Continents* Vavilov wrote:

> The account of the American experience with plant introduction furnishes much information; but at the same time it is evident that the effort was not based on a single dominating principle which is necessary for such a field of research, such as a hypothesis about phytogeography, the evolution of the plant kingdom, the succession of stages, the variability in space and over time and the peculiarities of cultivated and wild plants (Vavilov 1987b, p. 79).

During his three to four months in the USA, Vavilov visited plant research and crop production institutions in Maryland, Virginia, North and South Carolina, Kentucky, Indiana, Illinois, Iowa, Wisconsin, Minnesota, North and South Dakota, Wyoming, Colorado, Arizona, California, Oregon and Maine. Vavilov learned in great detail about American agricultural science, studying the seeds of the most popular cultivars of cereals, vegetables, melons, industrial and other crops. He established numerous business, scientific and personal contacts, which undoubtedly helped him to implement his programme of collecting and purchasing cultivar seeds, founding a seed collection, selecting literature and acquiring necessary equipment for the Department of Applied Botany and Breeding. During this trip Vavilov met L. Burbank. This meeting and what he had seen in his visit to this world-famous horticulturist impressed Vavilov very much.

In view of the problems caused by the drought of 1921 and the need to purchase seeds in the USA and Canada, Vavilov appointed D.N. Borodin, a Russian agronomist who had been living in America for a long time, as provisional representative of the Department of Applied Botany and Breeding. During this period the New York Division of the Department of Applied Botany and Plant Breeding gathered from all the States of America and partly from other countries a huge number of plant varieties, (about 20 000) and shipped them to Russia, collected numerous publications from all experimental stations, and established contacts not only with the USA, but also with other countries. In the full sense, it played the role of a window into the world for Russian experimental and agricultural institutions. Vavilov attributed great importance to the activities of this Division of his Department and always assisted it in receiving money transfers from Russia as, for the most, part the Division's work was maintained by the enthusiasm of its head officer Borodin and his voluntary assistants. Borodin himself provided weighty support to many Soviet scientists of that time who came to the USA, and helped them to establish contacts with American researchers and also to collect samples of cultivated plants. The American Division of the Department was referred to for the last time in Vavilov's letter dated 22 March 1924:

> With the purpose of communicating with various countries, the Department has been operating since 1921 a New York Division, opened by permission of the Department of Agriculture in Washington D.C. Through this Division it is acquiring numerous seed samples and literature which enables the Department to be aware of the latest research achievements of Western Europe and America (Vavilov 1980, p.157).

Thus, this Division of the Department of Applied Botany existed for about three and a half years till March 1925 and contributed a lot to the restoration of the destroyed relations between the scientists of both countries.

While returning from America, Vavilov visited many scientific research institutions in Europe. In the UK he met his teacher Bateson, Punnett, Percival and Beaven as well as other scientists. Bateson received from him a revised English manuscript of the law of homologous series for publication in the *Journal of Genetics*. After his trip to England and meeting with Bateson, Vavilov decided the ultimate direction of his future collecting missions as one of the approaches to solving the problem of origin of cultivated plants. In his letter from England Vavilov wrote:

> I spent three days with Bateson. We've talked about everything. A whole evening we've been talking about evolution. I guess, it was the most substantial discussion during all my trip...A journey to Africa becomes a necessity (Vavilov 1980, p. 43).

Fig. 8. New York Division of the Department of Applied Botany (1923). Sitting at the table: D.N. Borodin, Head of the Division [VIR Archives].

Fig. 9. New York Division of the Department of Applied Botany (1923), 136 Liberty Street, New York, NY. X = premises of the Division [VIR Archives].

In this letter Vavilov also wrote:

> I have visited Percival. I have seen the Abyssinian wheats and hope to receive about 200 samples of Afghani, Spanish and Portuguese wheat. If everything I have collected reaches us, our collection of cereals will obviously be the best in the world (Vavilov 1980, p. 43).

In France he visited the Vilmorin seed breeding and production company. In the Netherlands he had a meeting with de Vries in his laboratory near Amsterdam. In Germany he met Baur and Correns. In Sweden Vavilov became acquainted with the works of Nilsson-Ehle, which attracted his attention by "an ideal combination of agronomic expertise in agricultural conditions and requirements with a fundamental scientific background and geneticist's skill to capture the essence of each phenomenon" (Bakhteev 1988, p. 77).

After his seven months abroad (from July 1921 until February 1922) Vavilov came back to Petrograd overwhelmed with impressions, ideas and scientific information. The trip proved extremely useful and fruitful. It enabled him to establish numerous important contacts with his colleagues in America and Europe. Soon after his return, the Department of Applied Botany and Plant Breeding began receiving parcels with books, journals, reprints and seeds, along with the equipment purchased to meet the needs of plant genetics, cytology, physiology and evaluation of milling and baking quality of wheat cultivars.

Rebuilding the department

After his return Vavilov continued the process of restructuring and enhancing the research activities of the Department and reinforcing its scientific staff by employing prominent researchers. Vavilov was attracting scientists on the national level to studying the materials collected by him in foreign countries. In these studies Vavilov was assisted by S.G. Navashin, a cytologist, and S.I. Zhegalov, a geneticist and plant breeder from Moscow, G.A. Levitsky, a cytogeneticist from Kiev, N.A. Maximov, a physiologist from Tbilisi and G.S. Zaitsev, a botanist and geneticist from Tashkent (the three latter scientists later took posts with the Department of Applied Botany and Breeding). To attract the best researchers of Russia during the last half of 1922, Vavilov began corresponding with prominent scientists, such as the plant breeders V.V. Talanov and V.E. Pisarev, the agronomist P.T. Klokov, the botanical taxonomist P.M. Zhukovsky, and others. Later all of them accepted his invitations and came to work at the Department (Vavilov 1980).

In 1922 Vavilov continued expanding the work at the breeding and genetics station in Detskoye Selo. In his letter to Bateson of 1922, Vavilov wrote:

> Much of my time is taken up with the organisation of our new experimental station in the environs of Petrograd. You will be rather astonished to hear that the country-house we live in was many years ago presented by the late Queen Victoria to her godson the ex-Grand Duke Boris Vladimirovich. The country seat is very charming, the house itself quite English in style. During the past four years, unfortunately, the main buildings were occupied by comrades. (Vavilov 1922b).

In his letter to D.L. Rudzinsky of 1923, Vavilov wrote:

> In Tsarskoye Selo at the estate of Grand Duke Boris Vladimirovich we have established a station of good appearance thanks to the equipment imported from abroad ... (Vavilov 1980, p. 86).

In the extremely difficult situation of 1922, Vavilov, as the chief editor, revived *Works on Applied Botany and Plant Breeding*, suspended in 1917. The issues that had remained with the printer since 1917, including Regel's monograph *Cereals of Russia*, were published, as well as new issues in which Vavilov's personal contributions appeared, such as *On Recognizing Bread Wheats (Taxonomic and Geographic Essay)*, *Field Crops of the South-East* and others. Vavilov was also selecting from other regions of the country the best works in the sphere of botany, taxonomy, genetics, and other related disciplines for publishing them in subsequent issues of the *Works*. In 1922 Vavilov wrote:

> Already thrice I have cabled the Experimental Department (to Moscow) about our catastrophic situation with funds. We are unable to remunerate employees, hire day-workers, or pay for horses, and in general it is essentially impossible to work... (Vavilov 1980, p. 47).

> Never in their existence have the Department of Applied Botany and its sub-stations faced so difficult and unclear circumstances (Vavilov 1980, p. 46).

In a comparatively short time Vavilov overcame difficulties accompanying the process of creating the conditions for scientific research at the Department of Applied Botany and Plant Breeding in Petrograd at Detskoye Selo and at the provincial sites. At Detskoye Selo, studies of the ever-growing collections of cultivated plants resumed, and genetical investigations of those materials were also launched there. In his letter to Bateson at the end 1922, Vavilov wrote:

Fig. 10. Department of Applied Botany (1923) [VIR Archives].

Fig. 11. Department of Applied Botany (1923). On the table: materials of V.E. Pisarev's trips to Mongolia and the Far East [VIR Archives].

> By means of a detailed investigation of the cultivated plants of European and Asiatic Russia and the neighbouring countries, we endeavour to solve some problems of the origin of plants, making great use of the methods of genetics and cytology in our systematic investigations (Vavilov 1922b).

By this time the collected plants and seeds had been studied at the stations belonging to the Bureau, namely at the Steppe Experimental Station in Voronezh Governate and at the Moscow Division. The Turkestan Division opened at Tashkent State University, and the Murmansk Division of the Bureau.

In 1922, the State Institute of Experimental Agronomy (SIEA) was set up on the basis of the Agricultural Scientific Committee. The Department of Applied Botany and Breeding, headed by Vavilov (previously under the authority of the Committee), became part of the Institute along with other departments which represented independent institutes. In 1923, Vavilov was elected director of this Institute.

Geographically based studies of plant production

By autumn 1923, geographically oriented plantings were started with the purpose of studying ecogeographic regularities of ontogenesis in major cultivated plants. Permanent and temporary experimental stations and plots situated in contrasting environments were involved. The principles of such research had been laid in Regel's times with three stations located in contrasting climatic conditions. In 1923, in his letter to G.S. Zaitsev Vavilov wrote:

> This year the Department of Applied Botany and Plant Breeding of the Agricultural Scientific Committee initiates systematic research on the variability of the crop cultivars in different areas of the European and Asian parts of Russia. For the vast territories of European and Asian Russia the problem of morphological and physiological variations that the same cultivars undergo in different areas and under different conditions is of tremendous importance (Vavilov 1980, p.102).

Fig. 12. Institute of Experimental Agronomy (1923). Second from right: N.I. Vavilov [VIR Archives].

Initially, the plantings were organized at only 25 locations, but in 1927 the number increased to 115. The extreme locations of these plantings were near Murmansk in the north, Turkmenistan in the south, Kaunas in the west and Vladivostok in the east. Geographical experiments encompassed over 40 species of cultivated plants. All in all, 185 different spring and winter cultivars, represented mainly by pure lines, were planted from year to year according to a unified programme. Observations and maintenance of the plants were performed in compliance with respective instructions. In 1926, Vavilov wrote:

> The results of the last three years of our geographic experiments have revealed a number of regularities in variations of the chemical composition, morphological properties and vegetation period with the changes of longitudes and latitudes. In the current year, as in the previous one, the Institute has selected the most interesting groups of cultivars of different crops, including bread cereals and other field crops. The major Russian and foreign spring varieties have been included in the collection: for example, American cultivars and cereal races with disease resistance from Norway, Persia and Northern Africa – 170 samples in total (Vavilov 1980, p. 265).

While organizing these geographical plantings Vavilov first of all pursued the task of identifying possible geographical limits of agricultural crop distribution, so that he would be able to use the resulting conclusions as the basis of practical measures toward regulating the agricultural plantings in the country. It was essential to disclose regularities in the dependence of individual variability from geographic factors. In what way morphological and physiological traits and chemical composition of plants vary, what characters are conservative and consequently applicable for taxonomic purposes, what is the interrelation between the environments and heritability – these were the problems to be solved by means of geographical experiments.

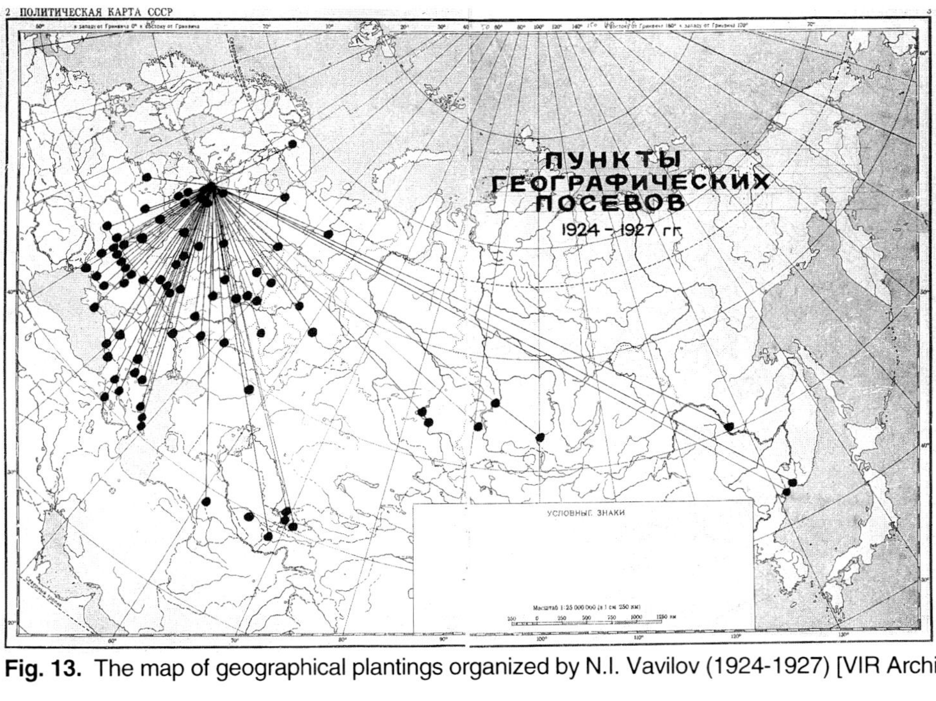

Fig. 13. The map of geographical plantings organized by N.I. Vavilov (1924-1927) [VIR Archives].

In 1924 Vavilov started organizing an amazingly extensive network of state trials of new and introduced varieties in the USSR by applying the same geographical principle which enabled him to replenish the results of geographical experiments. For this purpose a special Department of Variety Testing in the Institute was established, headed by prominent agronomist V.V. Talanov (Sinskaya 1991). The Department played an important role in introducing promising home and foreign varieties into wide cultivation.

The results of the first series of experiments were reported by Vavilov to the Scientific Board of the International Agronomic Institute in Rome. In his report Essais geographiques sur l'être de la variabilité des plantes cultivées en l'URSS (Russie) Vavilov (1927a) informed the world that the Institute under his leadership had commenced work on disclosing the laws of individual geographic variability, i.e. variations of the same genotypes under the effect of different geographical factors: longitude, latitude, etc. Concrete results of this research were presented by Vavilov to the conference participants. His report provoked an exceptional level of interest. A resolution was adopted concerning the necessity for geographical experiments with major bread cereals on an international scale on the basis of Vavilov's approaches.

In 1923 the All-Union Agricultural Exhibition was held in Moscow. This Exhibition served as a source for considerable replenishment of the crop collections with seeds of landraces from all over the USSR. Similar collecting had already been undertaken in Regel's time. As a result of the active and fruitful work of the Department, by 1924 the collections had grown up to 50 000 accessions thanks to exchanges by mail and direct collecting (Bakhteev 1988).

In 1924 Vavilov undertook an expedition to Afghanistan which resulted in adding approximately 7000 accessions to the collection. In the next year he travelled to Khoresm.

Vavilov's activities in the All-Union Institute

The foundation of the Institute and development of its work programme

In 1924, the Department of Applied Botany and Plant Breeding was rearranged into the All-Union Institute of Applied Botany and New Crops. In the same year the Presidium of the USSR Central Executive Committee adopted the statute concerning the organization of the V.I. Lenin All-Union Academy of Agricultural Sciences (VASKhNIL). Also, in 1925 the USSR Council of People's Commissars (Government) endorsed the Statutes of the All-Union Institute of Applied Botany and New Crops which formed the nucleus of the future Academy.

On the first session of the Institute's scientific board, which was attended by several members of the USSR Central Executive Committee, representatives of the Union Republics, the Russian Academy of Sciences, the State Institute of Experimental Agronomy and many other members of the scientific board, Vavilov presented a programme report entitled: "Consecutive Tasks of Agricultural Plant Science (Plant Resources of the Earth and Their Utilization)", in which he outlined the following objectives:

- Worldwide study of existing cultivated plants, registration of the plant materials available in this country and abroad that are ready to be utilized; wide-scale collection of plants and varieties at the areas of their origin; identification of the most valuable wild forms and their introduction into cultivation.
- Registration of plant varieties of all crops, organization of the State Variety Trials on the basis of planning, and definition of the boundaries for cultivated varieties.
- Utilization of wild plants for introducing new crops into cultivation.
- Mastering the synthesis of new plant forms that do not exist in natural environments (Vavilov 1990).

Later, all these tasks were transformed into the objectives of the Institute's activities and were implemented in its respective departments.

In a letter of February 1925 Vavilov formulated new tasks and aims for the Institute:

> The most valuable of all the possessions of the Department of Applied Botany – despite the great amount of work – is a large number of workers, among which, as you may be aware, there are quite a few prominent and unique specialists (such as Maltzev, Pisarev, Maximov, Pashkevich, Kichunov, Govorov and Flyaksberger); all of us are a consolidated unit that enables our ship to move forth towards her goal. We set high standards for ourselves: attempting to proceed forward, we are making use of all the best of the old traditions, for which the Department was renowned (Letters vol.1, p. 178).

The problem of the origin of cultivated plants retained its priority in practical collecting and organization of the Institute's collecting missions. In 1925, in a letter to G.D. Karpechenko, Vavilov wrote:

> For a number of cultivated plants, we may assume that we have got close to the centres of origin, and are facing the far more difficult problems of Linnean species formation. Several years later, provided we are still alive, the mapping of the major cultivated plants on the globe will be completed... (Vavilov 1980, p. 249).

In 1926-1927 Vavilov undertook a complex collecting mission to the Mediterranean countries, Abyssinia and Eritrea that resulted in the identification of the Abyssinian centre of origin of cultivated plants.

Fig. 14. K.A. Flyaksberger, researcher of the Department of Applied Botany responsible for wheat (1923) [VIR Archives].

Fig. 15. The first expanded session of the scientific board of the All-Union Institute of Applied Botany and New Crops (1925) in Moscow. In the centre: N.P. Gorbunov; to the right: N.I. Vavilov [VIR Archives].

Preliminary results of this expedition were formulated by Vavilov in his report entitled *Les centres mondiaux des genes du blé*, which he presented to the First International Wheat Congress that took place in Rome in April 1927, when he returned from an expedition to Russia. As Vavilov wrote:

> Abyssinia and Eritrea yielded materials of exceptional interest, far more interesting than those that I expected to find. The initial sources of durum wheats and barley have been reached... Abyssinia appeared to be an autonomous centre of development of the cultivated flora (Vavilov 1980, p. 296).

Right after Vavilov's report, participants of the Congress put forward a proposal to set up an experimental station meant for preserving that unique specific diversity.

Later in September 1927 at the Fifth International Genetic Congress in Berlin Vavilov in his report *Geographische Genzentren unserer Kulturpflanzen* summarized the data on other crops. In this report he put forward his idea that the dominant genes of any cultivated plant species were concentrated in the centre of its origin, while the recessive ones were commonly present and expressed at the periphery.

The achievements of Soviet genetics and plant breeding as well as the role of the Institute and its collections in those accomplishments were presented at the All-Union Congress on Genetics, Plant Breeding, Seed Production and Livestock Breeding held in January 1929 in Leningrad. Prominent foreign geneticists and plant breeders were present at the Congress, among whom were R. Goldschmidt and E. Baur from Germany and H. Federley and O. Valle from Finland. They unanimously acknowledged the extremely rapid progress of Soviet genetics and breeding. In his opening speech at the Congress,

Fig. 16. E. Baur and N.I. Vavilov (1929). 1st All-Union Congress on Genetics, Plant Breeding, Seed Production and Animal Farming, Leningrad [VIR Archives].

Fig. 17. Pavlovsk Experimental Station of the All-Union Institute of Applied Botany and New Crops (1926). Sitting in the centre: N.I. Kichunov; to his left: N.I. Vavilov; to his right: S.M.Bukasov; behind him: V.E. Pisarev [VIR Archives].

Vavilov remarked that Leningrad is the place where the science of genetics had originated and where Koelreuter had started his remarkable work on plant intercrossing 168 years ago. This Congress had three predecessors: the Congress on Breeding and Genetics in 1911 in Kharkov, the Congress in 1912 in St. Petersburg and the Congress in 1920 in Saratov. At the 1929 Congress Vavilov presented a paper entitled *The Problem of Origin of Cultivated Plants and Animals* (Vavilov 1980).

In mid-1929, Vavilov organized a collecting mission to China, Japan and Korea. The next year, he explored in the USA and made an intermediate stop in the United Kingdom. At the fifth International Botanical Congress in Cambridge, Vavilov presented a paper on one of his fundamental concepts "Linnean species as a system" where he presented the principles of investigating populations of cultivated plants, i.e experimental plant taxonomy, devised by him. In the Ninth International Horticultural Congress which took place at this time in London, Vavilov presented the results of fruit crop collecting accomplished by numerous missions within the USSR, and the study of materials collected at the Institute's stations. In his report *Wild Progenitors of the Fruit Trees of Turkistan and the Caucasus and the Problem of Origin of Fruit Trees,* Vavilov concentrated on the Caucasian focus of fruit tree and bush formation, the diversity of wild fruit plants in Middle Asia, wild fruit trees and shrubs of Siberia and the Far East, and the results and prospects of fruit tree research in general.

Participating at the Second International Conference of Agricultural Economists in Ithaca (USA) Vavilov gave a speech entitled "Science and Technology under Conditions of a Socialist Reconstruction of Agriculture". In the autumn of that year, he started implementing a plan of collecting missions in the southern states of the USA, Mexico, Guatemala and Honduras.

Genetical research

Following his collecting of plant resources in their centres of origin, studying their diversity in contrasting environments and understanding the taxonomy and evolution of species, Vavilov undertook an extensive programme on the genetics themselves proper of cultivated plants initiated in the 1920s at Detskoye Selo (now Pushkin). He formulated his understanding of this subject in his letters to G.D. Karpechenko, whom he invited to be the Head of the Institute's Department of Genetics:

> ...It is absolutely clear to us that the amount of work associated with genetic issues, that we are facing now, is tremendous. In practice, this work has already been launched in wheat, in barley, in oats and in the Cruciferae. The work in horticultural crops, melon and strawberry will come next... Genetic research should move towards solving the problem of experimental genetics and also be concentrated monographically on individual plant (species), on separate traits and, in particular, on interspecific hybridization (Vavilov 1980, p.258).

Vavilov then put forward even broader tasks for the Institute's geneticists, intending to combine in them the width of a system approach with depth of genetic research in studying the ever-growing collections of cultivated plants. He wrote:

> I must say that I am not at all satisfied with the current state of affairs in special genetics... I am still deeply interested in barley. Maybe, there will be as many samples as there should be, but to me, a taxonomist and geographer unable to neglect the questions of evolution, all we have looks like tiny fragments. I have not a trace of doubt that the characters found to have certain links in separate cases, would never be found in different combinations. The existence of links is a definite fact, but in this particular

case they are far from being clear. Apparently, a really colossal amount of material (land races, cultivars, forms, botanical varieties, etc.) is required in order to catch a regularity. However, we need to examine the whole genus (Vavilov 1987a, p.75).

The existence of a link between Albina and Clorina is just a tiny fact within the notion of the whole system of barley. The question of questions, of deepest interest for me is that of the interrelations between the Abyssinians, Daghestanians, east-Asians, and the wild species inhabiting south-west Asia. East-Asia has really separated out some amazing groups...

What is deemed necessary to me, is a real attempt at monographic treatment of the barley genetics considering various groups and at least a selective recording of characters for different geographical groups. In short, it is desirable to proceed from fragments and approach something as a whole that would deepen the knowledge of the intraspecific differentiation and help us to understand the process of mophological type-formation. I am interested particularly in the genetics of the east-Asiatic group in comparison with the barleys from south-west Asia and Abyssinia...

It is as clear as day that there is enough fascinating work in the sphere of genetics proper for decades for dozens of people, and it may lead up to general conclusions, as it is noteworthy that these ecotypes can be encountered in practically all plant (species): Abyssinian flax, Abyssinian barley, Abyssinian wheat, sesame – they are all saturated with anthocyanin; in Arabia Felix the earliest wheat in the world has been found, and the earliest barley has also been found there. Just recently I discovered that the blue-flowered alfalfa in Yemen has turned from a perennial into an annual plant. These subjects are worth considering from the genetical point of view, and there are scores of such subjects...

My observations of what is being done by geneticists make me think that, maybe, a combination of taxonomy with genetics will be especially important for this purpose. If specialists in general genetics expressed a wish to work on genetics proper, they would have been most welcome. There's enough of it for a whole generation, for hundreds of researchers, but the engine has to be started (Vavilov 1987a, p. 77).

So Vavilov regarded the problem of genetic research on cultivated plants and on each separate species in correlation with the intraspecific diversity, leading to geographical principles, which had always been the central basis of his research activities, as a keystone. By studying the genetics of separate species he strived to get a comprehesive overview of the hereditary nature of the whole species, which was possible only if such a study was performed not at random without any analysis of their origin, but on carefully made selections representing the varietal and botanical diversity of a given species. Therefore selection or identification and collection, if available, of the whole intraspecific or sometimes even intrageneric diversity was accepted by Vavilov as the first stage of any serious research on a plant species.

The results of the cultivated crop research activities performed by the Institute of Plant Industry in the 1920s were reported by Vavilov at the Second International Congress on the History of Science and Technology which was held in 1931 in London, UK. He gave a presentation entitled "The problem concerning the origin of agriculture in the light of recent research".

During this trip Vavilov used to his advantage visits to several scientific institutions, including Professor Percival's laboratory, which he mentioned in his letter to

O.K. Fortunatova dated 21 July 1931:

> ... I have just returned from London where I visited Percival. He continues working with the same persistence. It is wonderful to see that a 70-year-old man is still determined to move forward towards the chosen target. He performs many crosses with *Aegilops*... (Vavilov 1980, p.124).

Crop evolution and taxonomic studies

In September 1931, Vavilov was invited to visit Denmark and Sweden by the scientific communities of these countries to lecture and study the works of biological, plant breeding and agricultural research institutes. He visited Copenhagen, Lund, Stockholm, Svalof, Weibullsholme and Landskrona. Professor O. Gustavson recalled his meetings with Vavilov with great warmth. After this trip, Vavilov described his impressions in several letters to his colleagues working at the stations:

> I've been in Denmark and enjoyed their work in the struggle to put new lands into cultivation (Vavilov 1987a, p. 133)

> ...This year I have twice had a chance to go abroad lecturing in Denmark and Sweden. It was a great hindrance, although it was interesting because I met all the European plant breeders (Vavilov 1987a, p.144)

When summarizing the results of crop evolution research, Vavilov dealt with this problem in his report *The process of evolution in cultivated plants* at the Sixth International Genetical Congress which was held in Ithaca (USA) in August 1932. On the basis of the results of collecting missions of Vavilov himself and his colleagues he mused on the problem of the origin of cultivated plants. He said:

> Several regions of the globe are extremely rich at the present time in numbers of species and varieties of plants and animals. In North and Central America such regions are Southern Mexico, Guatemala and some adjacent small countries to the south. In Europe such regions are the Caucasus, the Balkan States, Italy and Spain. Huge areas in Northern and Central Asia are on the contrary quite poor in numbers of species. On the other hand, Southeastern China, India, Indo-China and the mountainous regions of Persia. Afghanistan, Russian Turkestan and Asia Minor are extremely rich in the variety of existing species of wild animals and plants (Vavilov 1932, p. 331).

Further on, he reviewed the complex nature of Linnaean species and the geographical principles of evolution that were derived from his study of the materials obtained during his collecting missions. On the basis of the law of homologous series in variation Vavilov demonstrated parallelism between different species and genera and discussed the role of natural selection and man in the evolution of cultivated plants. In conclusion, Vavilov said:

> In order to understand evolution and to guide our breeding work scientifically, even in application to our principal crops such as maize, wheat and cotton, we must go to the oldest agricultural countries, where the keys to the understanding of evolution are hidden (Vavilov 1932, p. 342).

Fig. 18. N.I. Vavilov, T. Morgan and N.V. Timofeyev-Ressovsky (1932) at the VI Genetical Congress in Ithaca, USA [VIR Archives].

After the Congress Vavilov visited the southern Canadian provinces of Ontario, Saskatchewan, Alberta and British Columbia and in the USA the states of Washington, Montana, Colorado and Kansas. Being in Canada and the USA, Vavilov paid special attention to the problems of irrigation in agriculture and noticed the valuable experience of these countries in crop rotation on irrigated farms, which could be applied in the USSR. Vavilov described his impressions of this trip in his letters:

> I've studied resistance having passed in three days the whole course as a laboratory assistant in Winnipeg...It was a surprise to know and see that the largest irrigated wheat-growing area is in Canada... With genetics, we are moving in the right direction: geographical crossings and emphasis upon interspecific hybridization. Here they lack good physiologists. Serious problems are avoided... We're lagging behind with milling... (Vavilov 1980, p.183)

> I have become well grounded in the irrigation of cereals, disease resistance and genetics (Vavilov 1987a, p.184).

In addition to the United States and Canada, Vavilov visited Cuba, Yucatan (Mexico), Peru, Bolivia, Chile, Brazil, Argentina, Uruguay, Trinidad and Puerto Rico to collect plants. After returning from America to Europe in February 1933, at an invitation from the French Society For Scientific Rapprochement with the Soviet Union, Vavilov delivered three lectures in Paris: on the results of his American expedition, on the agricultural science and land cultivation in the USSR and on the problem of origin of cultivated plants. In Halle (Germany), at the invitation of the Deutsche Akademie der Naturforscher Leopoldina, of which Vavilov had been made a corresponding member in 1929, he presented a report dedicated to the problems of plant science in the USSR in the socialist reconstruction period, and gave a lecture on "The Problem of Origin of Cultivated Plants", which was published in the Proceedings of this Academy in 1933.

Beginning in the autumn of 1932, Vavilov started a second broad series of geographical experiments. For such purposes the Institute's special Bureau for the Geography of Cultivated Plants was established. In this study were included different field and vegetable crops, so that tuber crops, root crops, berries and fruits were included. The work was carried out on a smaller number of sites than in the first cycle, but the studies involved a larger number of samples of each crop. As in the first cycle of experiments, the programme was based upon phenological observations, disease resistance assessments and chemistry. The variability of a number of morphological traits was also investigated. The last link in the chain of geographic experiments was the organization of cyclic crosses in cereals, legumes and flax. This programme was developed and partly implemented with Vavilov's personal participation. Such a thorough and fundamental programme of global hybridization in its geographic aspect could have been fulfilled only on such a diversity of materials as at that time was available at the Institute. The results of those studies disclosed extremely interesting regularities in the dominance of morphological and commercial traits, revealed valuable data concerning crossability, and helped to obtain new plant forms promising for breeding purposes. Unfortunately, this work was terminated at the very beginning of the 1940s.

Fundamental scientific publications

Such an extensive and detailed study of the world collection led to the identification of new regularities in the geographic variability of different characters and traits in cultivated plants. The results obtained were widely used by Vavilov and his collaborators in the volumes of the *Cultivated Flora of the USSR, Biochemistry of Cultivated Plants, Theoretical Bases of Plant Breeding* and *Cultivated Plant Trials,* as well as in dozens of monographs and hundreds of research publication (for example *Wild and Cultivated Oats* by A.I. Maltzev, 1930). In the mid-1930s, the first volumes of the *Cultivated Flora of the USSR,* a fundamental series Vavilov had been dreaming of in the 1920s, were published under his editorship. In a letter in 1933 Vavilov wrote:

> The All-Union Institute of Plant Industry has currently started compiling and publishing a voluminous work entitled "Cultivated Flora" which is meant to be a comprehensive botanical review of all cultivated plants which are, or can be cultivated in the USSR (Vavilov 1987a, p.188).

This series was intended to sum up the results of botanical collecting and studying of the world collections of cultivated plants and their relatives carried out by VIR researchers. When justifying the necessity for this publication, Vavilov argued in 1939:

> "The Cultivated Flora of the USSR" being published by the All-Union Institute of Plant Industry and Selkhoz Press is unique in the world literature as a publication devoted to the results of an all-embracing study of cultivated plants. It contains a taxonomy of cultivated species developed down to separate varieties, which is absent in all other Floras that encompass cultivated plants. It gives a plant breeder an opportunity to identify the plants he is working with, as well as to identify forms and varieties. In addition, the "Cultivated Flora" gives detailed references for each species concerning its geographical distribution, history of research on its biology and ecology, origin and

history of domestication, areas of cultivation in the USSR and abroad, economic value, variability, and breeding. Each entry for a species is supplied with a general bibliography grouped in compliance with the above aspects. The "Cultivated Flora" is a product of the joint effort of more than 80 experts on various crops, and therefore each separate chapter is not a compilation, but a collection of results of original experiments carried out for many years in the USSR. Thus, the "Cultivated Flora" in essence, is an encyclopaedia of cultivated plants, the major manual and reference book for each breeder...

Various cultivated species are classified in the volumes, not in compliance with the botanical order as it is usually done in other Floras, but according to their agricultural value. As a result, each volume of the Flora is a complete whole. As shown in the Plan of Publications, the "Cultivated Flora" comprises 22 volumes (Vavilov 1987a, p. 412).

From 1935 to 1941, the following volumes were published:

1. Grain cereals: Wheat
2. Grain cereals: Rye, Barley, Oats
4. Grain legumes
5. Fibre Crops. Part 1
7. Oleiferous Plants
16. Small Fruit Crops
17. Nut-Bearing Crops.

On the one hand, the volumes of the cultivated *Flora* were reviewing historical and modern achievements in the theory and practice of plant breeding, presenting a very detailed list of references on this problem, while on the other hand, they were monographic works on the taxonomy of an individual crop with a very detailed description of the area of its distribution and ways of utilization, including some considerations on the importance and adaptability of the genus and its species for environments in the USSR. Thus, volumes of this comprehensive publication could rightfully be called a *Cultivated Flora of the USSR*. They served as handbooks for plant breeders and botanists for many years, while some of them still remain unique monographs today. The volumes of this publication prepared under Vavilov's personal guidance, which had however failed to see the light of day during his life, were published in the 1950s (see Chapter 7). Simultaneously, preparations for publishing a fundamental manual entitled *Theoretical Bases of Plant Breeding* had been started. The first two volumes went to press in 1935, and the third one in 1937. Their total volume exceeds 2600 pages. In 1934, Vavilov wrote in a letter to D.L. Rudzinsky:

A gigantic work is being published, around 2 thousand pages, the "Theory of Plant Breeding" in 3 volumes, a joint work involving up to 60 authors (Vavilov 1987a, p. 241).

In the foreword to this edition, Vavilov wrote that the purpose of the book was to provide a critical review of the knowledge accumulated in the world regarding plant breeding and genetics. This collective effort of the researchers of the All-Union Institute of Plant Industry was based on the extremely important new factual materials obtained through studying the cultivated plants of the globe from the point of view of breeding objectives.

As already noted, the work was divided into three volumes. The first volume was dedicated to general issues and methods of plant breeding: botanical and geographical bases of plant breeding, principles of cultivated plant taxonomy, concept of mutations, theory of remote and intraspecific hybridization, and the application of cytological, anatomical, biometrical, biochemical and physiological techniques in breeding practice. The second and the third volumes addressed the breeding process in major cultivated plants. Particular attention was paid there to cereal grasses, fodder crops, industrial and fruit crops, vegetables and melons. The chapters dedicated to separate crops presented the results of the most recent crop research based on the study and introduction into breeding practice of the diverse varietal potential of plant species, found by the collecting missions of the Institute of Plant Industry.

As noted by Vavilov in the introduction, the collective effort made an attempt to sum up the level of modern knowledge and set further research objectives. The first two volumes contained several pieces by Vavilov, namely, Plant breeding as a science, Phytogeographical bases of plant breeding, The Law of homologous series in variation, The study of the immunity of plants to infectious diseases and Scientific bases of wheat breeding (Vavilov 1951).

Agro-ecological and taxonomic studies

Vavilov intended to present the basic principles of and approaches to the development of such a classification in a multi-volume publication of *World resources of cereals, grain leguminous crops and flax and their utilization in plant breeding*, of which only the first part was written by him in 1940 under the title *Agroecological survey of the principal field crops*. It was only in 1957 that this work was published (Vavilov 1957a). Putting forward arguments in favour of this kind of research, Vavilov wrote:

> The demands set by modern breeding practice, which has widely used materials for hybridization, call for the establishing of a new agroecological classification of existing varieties, that should be founded, in addition to the botanical taxonomy, on ecological, physiological and economically valuable traits and properties (Vavilov 1957a, p. 31).

Furthermore, in this publication Vavilov stated that regular research on numerous crop varieties collected in different countries under various conditions revealed the existence of regularities in their differentiation into agro-ecological groups. This was especially obvious when species widely distributed over the globe were cultivated in different, even contrasting environments. Such ecotypes were expressed in cultivated plants to no less an extent than in wild species.

Vavilov subdivided the major agricultural crops into certain ecotypes attributed to a certain agro-ecological area or region. In their turn, all the continents were divided into agro-ecological regions, which were subdivided into 95 corresponding areas. For example, for Europe he identified the following seven regions: Southern Europe, Mountainous areas of Western and Central Europe, North-Western Europe, European Steppe region, East European Forest-Steppe, Boreal region of East Europe and numerous regions of Siberia.

For the systematic study of all vegetation, Vavilov offered a list of major principles and properties to be taken into account when formulating agro-ecological classifications. Experimenting with geographic plantings allowed Vavilov to single out 29 parameters associated with plant habit, growth and development rate, different types of resistance to unfavourable factors and diseases, and technological properties of a given crop. Under Vavilov's guidance, VIR began extensive ecogeographical research that significantly increased our knowledge of cultivated plants in such aspects as morphology and taxonomy, as well as ecophysiology, cytogenetics and their practical application.

In his work entitled *New Systematics of Cultivated Plants*, published in English, Vavilov reviewed results achieved by the Institute and the first results on agro-ecological classification and wrote:

> In our study of cultivated plants we have advanced step by step. Some twenty years ago, coming to the study of cereals, we soon found that the previously existing classification into botanical varieties based on a few easily determined spike and kernel characters – which was elaborated by the German taxonomist Friedrich Koernike, and has been accepted by most investigators, including Professor John Percival in his monograph on the wheat plant (Percival 1921) – was not adequate. We found it necessary to elaborate a new, more detailed morphological and physiological system based on a study of the evolution of plants from their primary regions, which are usually characterized by the presence of a great diversity of botanical varieties. As a result of the establishment of the law of homologous series in variation (Vavilov 1922a), according to which closely allied species and genera to a great extent repeat one another in their differentiation, we discovered a huge number of previously known varieties unknown before. Many expeditions to the various primary regions of the origin of cultivated plants and thorough and many-sided studies of the collected material were conducted by the Institute of Plant Industry of the USSR. The evolutionary and geographical principle was taken as the chief basis of our studies of the species systems. We have tried as far as possible to follow in detail the steps of evolution from the primary regions where the differentiation into Linnean species took place. Fortunately, the location of these regions may be established on the basis of historical, archaeological and, particularly, botanical data (Vavilov 1940).

Vavilov mentions the discovery of many new species and varieties of cultivated plants:

> It suffices to mention that for wheat alone a dozen or so good new Linnean species have been discovered and hundreds of botanical varieties in the old botanical sense, each of which include many hereditary forms... Formerly there was known but one Linnean species of potatoes (*Solanum tuberosum* L.), but during the past decade Soviet expeditions have discovered, with the aid of cytologists, physiologists, and botanists, eighteen new species of cultivated potatoes and dozens of species of wild potatoes, some of them comprising many varieties (Vavilov 1940).

The needs of practical breeding led Vavilov to establish a new, agro-ecological classification of interspecific diversity on a global scale. For the period from 1930 to 1940 alone, the Institute of Plant Industry dispatched up to five million packages of seeds to research, breeding and other agricultural institutions, including those abroad. This voluminous material served as a basis for developing new varieties of cultivated plants for various climatic zones of the USSR (Bakhteev 1988). This work was performed by Vavilov by means of sowings in different geographical regions, i.e. studying the same varieties, collected in various parts of the world, under different conditions and testing the varieties' response to diseases and environmental factors. With this objective, experimental cyclic crossings were started with his personal participation, along with the development of agro-ecological classifications. Proceeding from the results of studies on these materials, Vavilov proposed to select the combinations that would be most promising in certain geographical zones of the USSR, and therefore accelerate the breeding performed by various research institutions. To this end, by 1940 a huge number of cyclic crosses had been carried out at research stations in Pushkin (the former Tzarskoye Selo), Kamennaya Steppe, Otrada

Kubanskaya, and at the Derbent Experiment Plot in Daghestan. Unfortunately these investigations were not completed.

Plant introduction and crop improvement

Vavilov wanted to help the development of agriculture in the most efficient way possible. He wanted all the breeders of the Soviet Union to use the VIR's world collections of cultivated plants more quickly and successfully. This aspect of research was dealt with by Vavilov in his paper entitled *Introduction of plants during the Soviet era and its results* (Vavilov 1940). He wrote:

> In contrast to the American introduction of plants, the Institute of Plant Industry is lately occupied not only with gathering material and reproducing it, but also with fitting it, as far as possible, into a strictly scientific system... (Vavilov 1965, p. 684).
>
> The subsequent stages, to which we have actually already turned with respect to a number of crops, consist of the selection of special, original material according to the revised theory concerning the selection of parents in conformity with various conditions for use during both intraspecific and distant hybridization. The present development of genetics, which has considerably broadened the plant-breeding basis to include distant species, and physiological research has, during our time, opened up new perspectives (Vavilov 1965, vol.5, p. 685).

Vavilov also mentioned the great practical work performed by the Institute of Plant Industry concerning the development and introduction of new varieties of agricultural crops:

> No less than 254 varieties have been put into production by the Institute itself and its research stations; one half of this consists of fruit and berry crops. Sixty-three crops are not included in this figure, since they were taken into cultivation after the introduction of material for the farming industry on the recommendation of the Institute and on the basis of research there. In addition, 52 varieties have already been introduced by other establishments, but based on collections obtained from the Institute of Plant Industry. More than 200 varieties are available from the Central Seed Testing Committee; many of these appear to have exceptionally good prospects. In all, out of the worldwide collection, the introduced varieties occupy more than two million hectares, in addition to the large amount of introduced varieties, mentioned above, which were obtained as a result of introductions for industrial purpose (Vavilov 1965, p. 686).

An important part of Vavilov's activities in the sphere of plant production was dedicated to the development of the extreme northern regions of the USSR. The Polar Experiment Station of VIR was established with this purpose in both scientific and practical aspects. This station coordinated the introduction of new plant species and varieties and worked out the methods of agricultural practice suitable for those extreme environments. In this regard, Vavilov was deeply interested in international experiences in northern agriculture, especially in Alaska, Canada, Iceland, Greenland and Scandinavia. In 1931, in his publication *The Problem of Northern Agriculture* (see Vavilov 1965) Vavilov wrote that in the process of its historical evolution the world's agricultural practice was moving northwards and down towards the tropics taking more and more space away from the forests. Both directions opened immeasurable expanses of uncultivated virgin lands for the agriculturist. Vavilov saw the future agriculture of the far north as an intensive type of agriculture, applying fertilizers, drainage, mechanization and electric power on a wide scale.

Much attention was paid by Vavilov to the development of a subtropical plant industry in his country. This is evident from the posthumously published work *Subtropical plant production in the USSR and its prospects* (Vavilov 1965). In it, Vavilov shows that the role of the All-Union Institute of Plant Industry in studying, collecting and introducing subtropical crops is tremendous. In Tzarist Russia, subtropical crops had been cultivated on about 1500 ha. By the mid-1930s, areas under these crops had increased to approximately 100 000 ha. The country had 50 000 ha under tea cultivation by early 1940, and tangerines, lemons and oranges on the Black Sea coast were grown on 17 000 ha. Land under tung cultivation (*Aleurites montana*) had reached over 16 000 ha. By the end of the 1930s, the cultivation of tropical and subtropical medicinal plants had been developed.

Under Vavilov's guidance, the problem of quinine cultivation in the USSR had been successfully solved. Thanks to joint efforts and the persistent labour of the VIR researchers at the Sukhumi and Batumi botanical gardens, a method for its cultivation had been developed. Some new crops had been introduced, namely several bamboo species, and many eucalyptus species that had spread widely along the Black Sea shore, especially in Abkhazia, Georgia and Adzharia, whilst the quinine plantations occupied up to 165 ha in Southern Transcaucasia. At this time, with the participation of researchers from VIR, a first inventory was made of the wild resources of subtropical plants.

The area of pistachio shrubs in Southern Turkmenia and Tadjikistan comprised 300 000 ha; the forests of walnut in Western Kirghizia, Southern Kazakhstan and mountainous Turkmenia covered tens of thousands of hectares; vast areas of pomegranate were found in Azerbaijan. Among wild almonds, walnuts and figs some forms were found that could compete with the best foreign cultivated varieties. In this connection, one of the immediate objectives for VIR and other institutions was to select (both within and outside the USSR) all that was valuable from the existing diversity of these important crops and introduce these materials into Azerbaijan, Central Asia and Kazakhstan, having increased the area under orchards tenfold. Special attention was paid to the development of filbert production in the southern foothills of the Main Caucasus Range. Vavilov wrote:

> A large volume of work has been carried out at the Introduction Nursery of the All-Union Institute of Plant Industry. Scientific grounds for introducing tung, quinine, eucalyptus, and many ornamental plants into culture have been developed... The Sochi Station maintains research aimed at the expansion of subtropical crops northwards. The contribution made by the Azerbaijan Station (network of VIR) in Mardakery to the planting of ornamental trees and bushes in the Apsheron peninsula and the development of subtropical crop cultivation should be noted... The Armenian Lenkoran Experiment Station (network of VIR) is working on developing the territory of Lenkoran for subtropical crop cultivation, and has had in these difficult conditions significant achievements with tea and subtropical fruit crops (Vavilov 1965, p. 697).

All the above-mentioned diverse interests of Vavilov materialized in the organization and activities of the All-Union Institute of Plant Industry. His encyclopedic knowledge promoted the development of new approaches to the generalization and classification of the results obtained, while his outstanding organizational talents made it possible for practical accomplishments to develop from the majority of his creative ideas.

Vavilov as a prominent organizer of science in the USSR

Beginning in 1925, Vavilov, who had become by that time a corresponding member of the USSR Academy of Sciences, participated very actively in the organization and establishment of the Academy of Agricultural Sciences. In 1926, being the Director of the All-Union Institute of Applied Botany and New Crops, Vavilov was elected to the USSR Central Executive Committee (TSIK). In the year when Vavilov's nomination as the VASKhNIL President was officially adopted he also became a member of the Collegia of the USSR Agricultural Commissariat.

Together with the previously established All-Union Institute of Applied Botany and New Crops that, in 1930, was renamed All-Union Institute of Plant Industry (VIR), 11 institutions were to be affiliated to VASKhNIL (Institutes of Agricultural Economics, Organization of Large-Scale Agriculture, Mechanization of Agriculture, Plant Disease and Pest Control, Drought Control, Land Reclamation, Land Cultivation, Livestock Breeding, Fisheries, and Maize) as well as the Basic Agricultural Library. By 1933, VASKhNIL incorporated 407 institutions, including 111 research institutes, 206 specialized zonal stations, 26 agricultural research stations, 36 breeding centres and 28 affiliates of central research institutes (Bakhteev 1988).

Thanks to Vavilov's initiative and with his active participation within VASKhNIL's network, the following research institutes were organized: the Institute of Grain Production of the South-East; Institute of Fruit Production; Institute of Vegetable Production; Institute of Subtropical Crops; Institute of Forages; Institute of Maize; Institute of Potatoes; Institute of Cotton Production; Institute of Flax; Institute of Oil-Producing Crops; Institute of Hemp; Institute of Soybean; Institute of Viticulture and Tea Growing, plus others (Soifer 1993).

At the same time as promoting the network of new institutions, VASKhNIL was responsible for training scientific personnel. For the first time in these years a very strong programme of postgraduate doctoral training was established and Vavilov was an active participant in this undertaking. By 1935, this postgraduate training service of the VASKhNIL institutes awarded more than 1300 doctorates.

In his report on the activities and reorganization of VASKhNIL in 1935, Vavilov pointed out that the control and, even more so, the management of such a cumbersome and ramified network of the most diverse institutions scattered all over the country, could not be satisfactory even with the Presidium working strenuously. Nevertheless, despite serious malfunctions, this period was marked by outstanding events not only of an organizational nature, but also in the development of the very essence of agricultural science (Vavilov 1980).

Until the mid-1930s, with Vavilov's participation, the institutes within the network of VASKhNIL underwent a process of intensive development. Institutes of both an applied and theoretical nature were multiplying in the country together with a large-scale enhancement of scientific staff training and reinforcement of contacts with foreign research institutions. In 1935, Vavilov was elected a full member of VASKhNIL and concurrently was released from the VASKhNIL President's position, discharging the office of its Vice President until August 1940. In 1938, Trofim Lysenko became the President of the Academy. In spite of the reorganization of VASKhNIL in 1935, its functioning continued to lack efficiency. However, with Lysenko coming to the leading position the Academy was transformed into an almost purely administrative authority (Bakhteev 1988).

In 1930, Vavilov was elected as a full member of the USSR Academy of Sciences, and after the death of Yu. A. Filipchenko, assumed responsibility for the Laboratory of Genetics of the Academy of Sciences of the USSR. For the decade to follow, he was its most energetic leader. At the end of 1933, the Plenary Session of the Academy of Sciences resolved to reorganize the Laboratory into the Institute of Genetics. As a consequence, the range of Vavilov's scientific activities increased significantly. When justifying the necessity of transforming the Laboratory of Genetics into an institute, Vavilov reported to the Presidium of the Academy of Sciences of the USSR:

> The framework of a modest Laboratory of Genetics cannot now be accepted by the Academy as satisfactory. In fact, the Laboratory has long ago outgrown its premises having expanded with glasshouses and fields; a considerable part of the work has been done by the Laboratory in various regions. The Laboratory has organized repeated expeditions to Central Asian republics, and to Mongolia. It possesses a glass-house of its own and a territory near the Botanical Gardens. In the near future we are expecting to have test fields to launch experimental work there. Many prominent foreign geneticists, for instance Dr Bridges, Dr Kostov, Dr Meller, have been working continuously at the Laboratory... I ask you to consider this question and approve the reorganization of the Laboratory of Genetics into an Institute starting from January 1, 1934 (Vavilov 1987a, p.210).

In his report at a Session of the Academy of Sciences of the USSR in February 1934, Vavilov singled out five major problems on which the Institute of Genetics was concentrating, namely

1. development of the theory of mutations and the related problem of the gene
2. interspecific hybridization
3. chromosomal studies
4. inheritance of quantitative characters
5. origin of domestic animals and cultivated plants.

Along with the scientists from this country such as A.A. Sapegin, G.A. Levitsky, A.A. Shmuk, T.K. Lepin, Ya.Ya. Luss and others, foreign experts such as K. Bridges, G. Meller and Doncho Kostov were invited to work at the Institute. In 1930 F.G. Dobzhansky from the USA and in 1933 N.V.Timofeev-Resovsky from Germany had been invited by Vavilov to work in the Institute (Vavilov 1987a).

G. Meller, together with Dr Offerman, a geneticist from Argentina, considerably accelerated the development of the material basis for research on mutations. Young scientists and doctoral students were involved in this work. Doncho Kostov, a prominent geneticist and cytologist from Bulgaria, worked on the interspecific hybridization of tobacco and G.A. Levitsky guided the research work on chromosomes. Thanks to the application of new methods, he managed to reveal details of chromosome structure. A study of quantitative characters in wheat launched with the participation of Yu. A. Filipchenko was continued by T.K. Lepin. By the end of the 1930s, wide-scale research on cyclic crosses of cereals, grain legumes, and flax had been launched jointly with VIR.

The public scientific activities of Vavilov were also extensive. From 1931 until the end, Vavilov was the President of the Geographic Society of the USSR. He contributed much to vitalizing the Society's work by helping with the publishing of its papers, providing for the expansion of its library and making personal presentations after each collecting mission.

Vavilov also participated in a number of other public scientific organizations, being a member of the Presidium of the All-Union Association for Oriental Studies, member of the All-Union Botanical Society, honorary member of the Moscow Society of Natural Science Experimenters, and many others.

Fig. 19. G.A. Levitsky, Head of the Department of Cytology at VIR (1933) [VIR Archives].

Fig. 20. K. Bridges addressing the extraordinary session of the USSR Academy of Science in Leningrad (1931). Second from the right, sitting: N.I. Vavilov [VIR Archives].

Owing to his ever-growing renown among the foreign scientific community, Vavilov was elected an honorary member of the British Association of Biologists in 1929, honorary member of the British Horticultural Society in 1930, foreign member of the Czechoslovak Academy of Agricultural Sciences in 1931, Doctor Emeritus of the Higher Agricultural School in Brno (Czechoslovakia) and full member of the Czechoslovak Academy of Sciences in 1936, and honorary member of the Indian Academy of Sciences and the Royal Scottish Academy of Sciences in the following year. In addition to these titles, Vavilov was elected to membership of the Royal Society and the Linnean Society in London, the New York Geographic Society, the American Botanical Society, the Mexican Agronomic Society, the Spanish Society of Natural Explorers, the Cyril and Methodius Society in Bulgaria, and made a Doctorate Emeritus of the Sofia University, Bulgaria (Zhukovsky 1967).

The tireless activities of Vavilov, as VASKhNIL President, Director of the All-Union Institute of Plant Industry and Director of the Institute of Genetics of the USSR Academy of Sciences, had an impact on the development of biological and agricultural sciences in the Soviet Union and provided for the growth of their prestige both in the USSR and abroad.

Vavilov and the Institute's international relationships, 1925-1938

International contacts with numerous organizations and experts in various fields of biological science made up a significant part of the Institute's activities. As Vavilov and his researchers made foreign trips aimed at collecting plant materials and gathering scientific information, so foreign experts visited the Institute in Leningrad and various regions in the USSR to collect plant genetic resources of wild and cultivated species, as well as to get acquainted with various aspects of the scientific activities of Vavilov, the Institute and its stations.

British and American visitors

In 1925 W. Bateson visited Leningrad and Moscow at the invitation of the Academy of Sciences of the USSR (he had been elected a corresponding member of the Academy in 1924) to attend the Two Hundredth Centennial of the Russian Academy. While in Leningrad, Bateson visited Vavilov's Institute and its experimental fields in Detskoye Selo, and became acquainted with the work of the Laboratory of Zoological and Botanical Research under Professor Filipchenko. In Moscow, he familiarized himself with Professor Navashin's cytological research, visited the Institute of Experimental Biology under Professor Koltsov and Professor Serebrovsky, and paid a visit to Professor Prianishnikov at the Agricultural School at Petrovsky-Razumovsky. Bateson's impressions of the visit to the Institute in Leningrad were expressed in the article *Science in Russia* published in *Nature* in 1925:

> The work is in the hands of Professor Vavilov, who has already built up a great establishment for this purpose, employing 350 people, of whom some 200 are trained workers. In his travels through Turkestan, Afghanistan and neighbouring countries and by a vast correspondence, collections of seeds of wheat, barley, rye, millet, flax, etc., have been brought together on a great scale. The central office is in Leningrad and occupies a very large building, in great measure a living museum of economic plants as

Fig. 21. At the celebration of the 200th anniversary of the USSR Academy of Sciences (1925). Biological Institute of Leningrad State University, Peterhof near Leningrad. [VIR Archives].

> represented by their seeds. Of wheat alone, some 13 000 forms are collected here. In various parts of the country there are twelve subordinate stations, and by sowings made once in about three years it is proposed to perpetuate most of the collection alive. Besides a cytological department under Professor Levitzky, there are special sections for meteorology, statistics, etc. Exceptional opportunities have been provided for investigating the geographical distribution of cultivated plants, especially in its bearing on problems of origin, which have resulted in several novel suggestions. As Detskoye Selo is the home breeding-station of this Institute. The chief building on it is a pleasant house, originally intended as a villa for Queen Victoria,[3] and adjacent to this a number of laboratories and additional accommodation have been provided (Bateson 1925).

Traditionally, the territory of Tzarist Russia and then the USSR had been visited by American researchers from the Bureau of Plant Industry, US Department of Agriculture in pursuit of forage, fruit and vegetable crops. At the initiative of the Bureau of Plant Industry, a collecting mission was sent to the USSR in 1929 to explore Turkestan, and to collect bacterial wilt resistant varieties and populations of alfalfa, as well as wild fruit and garden crops.

Two experts from the Bureau of Plant Industry – Dr Harvey L. Westover, Senior Agronomist, Alfalfa Investigations, and Dr W.E. Whitehouse participated in the collecting mission of 1929. This mission experienced great difficulties, as the USSR and the USA had no diplomatic relations at that time. However, having finally overcome all problems in early July 1929, the American experts arrived in Leningrad. Dr Westover described it in his report as follows:

---

[3] Actually provided by Queen Victoria for her godson the Grand Duke Boris Vladimirovich.

Fig. 22. At the celebration of the 200th anniversary of the USSR Academy of Sciences (1925). Laboratories of the All-Union Institute of Applied Botany and New Crops, Detskoye Selo near Leningrad (former estate of Grand Duke Boris Vladimirovich). In the front to the right: W. Bateson; to the left among his colleagues: N.I. Vavilov [VIR Archives].

> Three or four days were spent in Leningrad, during which time we visited the Institute of Applied Botany, the Herbarium, the Botanical Garden, the Institute of Plant Pathology, and the experimental Stations and Laboratories at Detskoye Selo. Quite a number of employees of the Institute of Applied Botany speak English and explained their work in detail (Westover 1929).

Throughout the trip, the American experts were accompanied by Professor Kuleshov. From Leningrad they moved to Moscow. Further, they went to Kiev, Kharkov, Valki (an experimental station near Kharkov), Saratov, Krasnyi Koot experimental station and Samara where they familiarized themselves with the work of the agricultural experimental station of the Cotton Committee stations. From Samara, the researchers departed to Tashkent. After Tashkent, the team moved to Frunze and then to Alma-Ata where seed-collecting in the neighbouring mountains was started. Westover further wrote:

> While at Alma-Ata we met Dr Vavilov, who had been on an expedition in western China, but was compelled to leave there as a result of the difficulties that had arisen between China and Russia (Westover 1929).

Fig. 23. Meeting of Vavilov with the USDA collecting mission on the way to Xinjiang (Alma Ata, Kazakhstan, 1929). Second from the left: N.N. Kuleshov, A.T. Belov, N.I. Vavilov and H.L.Westover [VIR Archives].

Fig. 24. At the Polar Experimental Station of VIR (1920s). N.I. Vavilov is standing in the centre [VIR Archives].

Then the team returned to Frunze where it continued collecting activities, and later it visited Tashkent where some of the nearby alfalfa seed producing districts were visited. Having completed the work at Tashkent the team split, and Dr Whitehouse together with his companions went toward Andizhan to collect seeds of wild pear and pistachio, while Dr Westover and Professor Kuleshov collected alfalfa seed along the road to Syr-Darya river, Golodnaya Steppe, Fergana and Samarkand. The two groups met again near Samarkand and jointly explored the regions around Bukhara and Chardzhou. The collecting mission intended to visit Tashauz and Prosu, the regions of special interest for the Americans, but "the chief military officer did not want us to proceed any further because bandits were giving much trouble in that region" (Westover 1929).

However, negotiations resulted in two soldiers joining the mission to guard it, and the team proceeded on the route. A large number of alfalfa samples were collected without any special complications. The team returned to Chardzhou and then took a train to Ashkhabad. From Ashkhabad, they proceeded westward by train to the Caspian Sea, which they crossed by boat to Baku, where they took another train for Sochi on the Black Sea. After spending a short time in this vicinity, they proceeded by boat to Sukhum, where there was a branch experimental station of the Institute of Applied Botany devoted primarily to the study of tropical fruits, aromatic plants, medicinal plants and forest trees. From Sukhum, they returned to Sochi by bus and then by train to Moscow. The mission was completed in early September 1929. A large amount of alfalfa seed samples, wild fruit and various vegetable and melon crops (the famous local melon varieties, in particular) was collected.

The next visit of American researchers from the Bureau of Plant Industry, US Department of Agriculture was that of Mr Morse and Mr Dorset, organized in 1930. In that year two phytopathologists – Professor L.R. Jones from Wisconsin University and Mr J. Dickson from USA – and the geneticist Dr Matsuura from Japan also visited the Institute of Plant Industry. In the 1930s the Soviet Union was visited by another USDA expert, Dr W. Dixon. One of his aims was to collect wheat and other cereals resistant to various rust strains. While in Britain and the USA in 1930, Vavilov invited a British fruit-grower named Benyard and an American pomologist (C. Swingle) to visit the USSR with a collecting mission to Central Asia and the Caucasus (Vavilov 1987a).

In 1932, Vavilov's Institute was visited by Dr T. Henry, a biologist from Argentina, who received a large set of seed material and some literature, and by Dr Mosley, an English biologist who undertook a trip to Siberia. When going to the USA in 1932, Vavilov was authorized by the Peoples' Commissioner for Agriculture to carry out negotiations with the experts whose visit to the USSR was desirable for consultations regarding some issues of agricultural biology and genetics. Vavilov invited the following experts: Dr J. Harland (Director of the British Imperial Cotton-Growing Experimental Station, Trinidad, prominent cotton breeder and geneticist, and the author of significant research programmes aimed at developing new methods of obtaining high-yielding intraspecific hybrids), Dr H.J. Muller (the most prominent geneticist of that time) and Dr H.W. Alberts (Director of the Alaskan Experimental Station who did much work in the sphere of northern agriculture). The latter even expressed a wish to get a permanent job in view of the closing down of the Alaskan Station due to the economic crisis, and Vavilov devised a plan to take Dr Alberts first to the Branch of the Institute in Khibini Mountains, and later to the Pechora river and other regions where intensive work on expanding agriculture further north was going on (Vavilov 1987a).

Professor Harland visited the Soviet Union in autumn 1933. In order to become acquainted with the status of cotton breeding and provide necessary consultations, Harland together with Vavilov made a trip to Baku (Azerbaijan) and toured Uzbekistan and other cotton-growing regions of the USSR where major institutions involved in cotton research and breeding were located. Theoretical assistance from Harland, his consultations with Vavilov, and the availability of a rich cotton collection brought by Vavilov from Egypt and South and Central America positively influenced the further development of cotton breeding and related research in the USSR. In 1934, Vavilov received a letter from Dr Harland expressing his willingness to come to the Soviet Union and get a permanent job there, but unfortunately these plans were never realized (Vavilov 1987a).

In the Autumn of 1933, Dr Muller came to the USSR to work at the Genetics Laboratory of the Academy of Sciences. In June and July of 1934 Vavilov, together with Dr Muller and a Dr Offerman, his assistant from Argentina, made a trip to the Caucasus. In his letter of 13 June 1934 Vavilov wrote:

> We are travelling from Rostov together with American professors, Dr Muller and Dr Offerman, and later with Dr Kostov in order to get acquainted with the work of research institutions in the Caucasus and Transcaucasia... (Vavilov 1987a, p. 236).

They visited Kiev, Rostov, Krasnodar, Baku, Gyandzha and Armenia. Dr Muller described this trip in a letter to M. Popovsky of 16 July 1966:

> It was a rare privilege when Nikolai Ivanovich took me and his son with him, along with an Armenian agronomist (Tumanyan), my American assistant (Offerman), and the chauffeur, on a long trip, mostly by car, through the Caucasus and Trans-Caucasus, where we visited his plant breeding stations. On the trip he also took us to see a youth hostel in the wilds, a collective farm and a state farm, a tractor works (in Kiev, Ukraine), and several attractive resorts. Once we went (or tried to go) by plane from Ganja to Baku, but over its airport were told that we must not land because there was a wind of 86 miles per hour. On turning back, the pilot informed us that we could not return either, having fuel for but a short distance. It was only natural that some of us were frightened – one in fact writing a will. On the contrary, Nikolai Ivanovich, to our amazement, since there was for once no way for him to help out, relaxed and caught up on some of his sleep. Fortunately the pilot managed to bring the plane down safely in a field, in the shelter of some low hills, near a little town and rail road, and we actually caught a train that got us to Baku that evening (Muller 1966).

This incident was by no means the only one on that trip which illustrated Nikolai Ivanovich's equanimity.

The next time reseachers of the Bureau of Plant Industry, US Department of Agriculture visited the USSR was in 1934. The American expedition with the participation of Dr Harvey L. Westover and Dr Charles R. Enlow explored the territory of Central Asia. In mid-May 1934, the two representatives of the Bureau of Plant Industry, USDA arrived in Moscow. In late May they visited Leningrad where Vavilov helped them solve several problems and expedited their departure. In his letter to Mr B.Y. Morrison, Westover wrote:

> Dr Vavilov has arranged for us to stay at his experimental stations and use such transportation as is available for trips... We have had a very interesting time in Leningrad visiting scientific institutes. Dr Vavilov and his staff have done everything they could to make our enforced delay interesting and worthwhile (Westover 1934).

In one of his letters to Dr Pieters, Westover shared his impressions of the activities of Vavilov's Institute:

> We spent two very interesting days at Detskoye Selo, where the Institute of Plant Industry is conducting extensive experiments. 'Yarovization' (which we would now call vernalization) was explained to us in much detail. We were told that several million hectares were sown to grain so treated, mostly spring wheat. It is claimed that the treatment increases the yield on the average 1¾ centners (175 kg) per hectare. Yarovization of winter wheat is more costly and the advantages are not sufficient to justify the additional cost. At present they are much interested in yarovized barley. From Leningrad northwards, barley seldom matures but where yarovized, good yields are obtained. Much work is being done along physiological lines, particularly in a study of wind burn and scald of wheat,and long and short day experiments. The breeding work being done under Dr Karpechenko is of special interest. By revegetation of cabbage they have succeeded in doubling the number of chromosomes, and seed from such plants produces much larger flowers, longer heads and plants that are in general more vigorous. Similar results have been accomplished with the rose geranium...
>
> In all breeding work special attention is given to its relation to geographical distribution of types. The origin and geographical distribution of cultivated plants are a sort of "hobby" with Dr Vavilov. I forgot to mention the work that is in progress here as well as at Kaizer Wilhelm Institute in Germany, to develop strains of lupins free from alkaloids. I don't know how the tests are made in Germany, but here the process is quite simple. With a sharp knife they scrape off a little of the seed and treat it with a solution of iodine. If the seed contains alkaloids the colour darkens, and if not, it is unchanged. In this way they have tested millions of plants in large fields and saved seed from those that do not contain alkaloids. They have many such plants and are now increasing the seed (Westover 1934).

The expedition managed to leave for Central Asia on 10 June. Its route was as follows: Ashkhabad – Dushanbe – Repetek – Alma-Ata – Tashkent – Frunze – Alma-Ata – Issyk-Kul – Chelkar. On 28 August the expedition returned to Moscow. The mission had collected over 1000 seed samples of grain cereals, grain legumes, forage and vegetable crops. In one of his last letters, Westover expressed his impression of Vavilov:

> Dr Vavilov has been of wonderful assistance during all our difficulties, and we wonder if it wouldn't be possible for the Secretary of Agriculture and Mr Ryerson (Chief of Bureau of Plant Industry) to write him letters of appreciation. I am sure this would please him. Friday morning the next day, I had a cable saying that Dr Vavilov wanted to see me the next day, as he was leaving Leningrad that night and would be away for some days. I caught the night train and spent a very interesting day with Dr Vavilov. He gave me some excellent advice regarding future expeditions. He is convinced that there is little of interest for us in Chinese Turkestan and says travel there is very difficult. He thinks Armenia offers a much more profitable field for exploration and thinks arrangements could be made to do some work there next summer if we care to (Westover 1934).

From Westover's letters it also became known that Professor Hansen, who had repeatedly visited Russia and the USSR, was in the Altai mountains with a collecting mission in 1934. That year, as Vavilov wrote in a letter to Vysotsky, the Institute:

> ...was flooded with foreign specialists, a number of prominent geneticists, in particular. Among others, C. Darlington, an English cytogeneticist, visited the Institute (Vavilov 1987a, p. 245).

Visits by O. Frankel and others

In 1935 the Institute was visited by Otto Frankel, a young breeder from New Zealand. He described the purposes of his trip to the USSR in a letter to Vavilov of 2 July 1935:

> My department (the New Zealand Department of Scientific and Industrial Research) wishes me to see as much as possible of your work on wheat and on pasture plants. But anything else in plant breeding, genetics and cytology will be of greatest interest. Although I should be anxious to see some of the wheat areas in the South, it would hardly be worthwhile, I believe, to travel too much during this short period (Frankel 1935).

Nowadays Sir Otto H. Frankel recollects this trip thus:

> I visited the Soviet Union in 1935, mainly to meet Vavilov. In that time I was a young and unknown scientist from New Zealand. Vavilov was welcoming and most generous with his time and his ideas. I was able to spend nearly all of my 6 or 7 days in Leningrad with him, either at the Institute or visiting research stations. I had an opportunity to visit the collection of potato in Detskoye Selo near Leningrad and the Institute's research station in Kharkov (Ukraine) and some places in Moscow and Kiev. In the Institute I met Bukasov – potato specialist, Flyaksberger – wheat specialist, and others. Vavilov was of middle height, broad-shouldered, very dark, an impressive but not obtrusive personality, unhurried yet very active mentally and physically – one of his constant phrases was "Let's go – life is short". He told me he never slept more than 2 hours; when he dropped me at my hotel late in the evening he usually had an armful of journals to look through during the night, at that time it was New Zealand Agricultural Journals (Frankel, pers. comm., 1994).

In August 1935 after this trip Vavilov answered Frankel's letter:

> Thank you for your friendly criticism and suggestions. We shall take them seriously. I have spoken already to our chemistry people. Recently I talked with some of the best European biochemists, such as Dr Adberhalden and Dr Bertrand, who were here at the Physiological Congress, and I believe we shall introduce some changes in our work (Vavilov 1935).

Not only foreign breeders and geneticists, but also historians of science showed interest in the work of Vavilov and his Institute. In one of his letters to Academician I.I. Meshchaninov, an archaeologist, Vavilov wrote in 1935:

> I am forwarding to you a young French scientist, Mr Audricour who's been sent on a scientific mission to Leningrad by the Ministry of Public Education. He is interested in the history of agriculture, origin of cultivated plants, and in linguistics in relation to agriculture. He has already worked at our Institute for eight months and visited the largest experimental institutions in the Caucasus and Central Asia. He is preparing for publication a work for the series "La géographie humaine", a book devoted to civilization and the history of cultivated plants. I kindly ask you to acquaint Mr Audricour with the work of the Academy on agricultural history, as well as with the linguistical aspect of agricultural names. He has launched a comparative study of the names of cultivated plants in various languages (Vavilov 1987a, p. 287).

In 1937, a Swedish agronomist, Evert Oberg, visited the Institute in order to familiarize himself with the works in the sphere of breeding and applied botany. Vavilov mentioned this visit in his letter to T.D. Lysenko:

> He (Oberg) has worked in Leningrad for 3 weeks on a special monograph on Tibetan barleys and wheats. He's been studying the results of a Swedish collecting mission to Tibet in 1934-1935. Of special interest to him are the works devoted to grain cereals... Oberg has several hours at his disposal, and it would be good if you acquaint him with your work for 2 or 3 hours (Vavilov 1987a, p. 345).

A visit by J.G. Hawkes

In 1938 from 28 August to 10 September Vavilov was visited by J.G. Hawkes from Cambridge, England. He recollects his visit to the Institute thus:

> My purpose was to study the results of the Russian potato collecting expedition to South America by N.I. Vavilov, S.M. Bukasov and S.V. Juzepczuk, and thus to plan the I.A.B. (Imperial Agricultural Bureaux) expedition 1938-39 in relation to the Russian work. My immediate impressions of Vavilov were that he was a man in a million, very jovial, and putting one at his ease immediately, with a most interesting conversation (Hawkes, pers. comm.,1995)

They discussed some of the issues illustrated by Vavilov's activities and of course they touched on questions concerning Lysenko:

> Vavilov considered Lysenko to be just lucky and intuitive – he is not a real scientist at all and certainly not a geneticist, having no conception at all of modern genetics or its terminology. Lysenko only recognises three authorities 1. Darwin 2. Burbank, in America and 3. Timiryazev in Russia. Everyone else is "bourgeois and anti-Darwinist". Lysenko has a marvellous knack of publicity and the government and people are on his side. He gets quick results at any rate... But Lysenko states that this shows that the whole of Mendelism and the chromosome theory of heredity are erroneous, and that any type of hybrid can be produced his way. The only thing that worries Vavilov is that Lysenko is getting more money for his work at the expense of Vavilov; Vavilov is always being harassed by the government for practical results...

Hawkes further writes:

> He (Vavilov) then showed me a copy of a periodical entitled "Vernalization" by Lysenko which is one big tirade against genetics. Lysenko regards the Mendelists and Morganists in the worst possible light as pseudo-scientists and Mendelism merely as an expression of averages naturally obtained. Vavilov is going to write a criticism of his papers but doubts whether he will be able to get it published.

Next morning Hawkes had breakfast with Vavilov and he

> showed me a short column in the papers about my arrival in Russia stating that "Dr J. Hawkes, vice director of the Imperial Bureau of Plant Genetics has come to Russia in the Felix Dzerjinsky to inspect Russian genetical research and has brought some valuable specimens of potato tubers for the Russian genetists".

We then drove to Pushkin town where the experimental fields of the Institute were established ... The first thing I was shown by Vavilov was his wheat collection from all over the world. Every conceivable type grows here. The collections from China are completely awnless and he thinks this may be a new species but has not yet described it. He is led to believe that in certain areas parallel or converging evolution has taken place; thus in China he also found a beardless barley. The phenotypes are strangely similar though the genotypes differ no doubt. Mutations which vary in every direction are selected by the particular environment so that similar phenotypic characters are produced from possibly very different gene complexes. This theoretical conclusion has great practical value. If, for instance we know that the beardless wheat occurs in China and we wish to discover a beardless variety of barley, then the chances of finding this in the same region are very great. This clearly cannot be due to any relationship between them but rather to the type of environment moulding the random mutations into one sort of plant. This holds for wild as well as cultivated plants. The question of primary centres of origin for cultivated plants is bound up with all the factors of civilisation, whether the largest evolutionary areas are conditioned by the presence of man or *vice versa* is not clear, probably each conditions the other. The different types of ecological habitats to be found in the primary centres of form origination are no doubt a very important factor in producing diversification but Vavilov agrees that the fixation of species is not due so much to the presence of geographical barriers but on the other hand barriers due to chromosomal changes, either differences in number, or in the arrangement of groups of centromeres due to fragmentation, translocation or interchange, preventing pairing at meiosis and isolating the forms by sterility barriers... (Hawkes, pers. comm., 1995).

About his last day in the Institute Hawkes writes:

Vavilov showed me some of the work that is going on at the Institute. He has more and more of my admiration – the rare combination of a practical man and theoretician. He has made complete ecological surveys of Russia for agricultural purposes and advised on the best crops to be grown. His world collection of plants – the seeds of which are stored in the Institute – is colossal, requiring a very large number of workers to keep the collection going and in a living condition. The seeds are sown according to the various habitats in any of his stations scattered over the country about every three years...

He hoped to be able to come to the International Conference of Genetics in 1939 and wanted me to tell the people in England, i.e. Haldane, Crew, etc. that if he appeared offhand it was not due to him but to difficulties with the goverment. The money question, he hastened to assure me, was quite all right as this year he was getting an even larger budget than before. But there is still some fighting going on between him and Lysenko all the time which makes relations a little strained, since Lysenko can do no wrong in the eyes of the government...

Thus when Vavilov asks for permission to leave for England there may be some opposition from the Lysenko quarter. I can realise now, I think, why the government thinks so highly of Lysenko; he represents not the man so much as the idea. Of common peasant stock he has risen to the highest place of honour in soviet intellectual life – that of Academician. Whether he is worth it or not is another matter. They are overwhelmed by the fact that under the new regime a man can reach the highest place from the lowest by virtue of his own efforts alone. That is a grand thing but I have a feeling it may be something of a wish fulfilment for the people who voted him there. Vavilov can lay hold to no such claims as this, for he had some of his education in England and America before the revolution. He does not claim, as Lysenko does, that Communism has done everything for him. He would have been great in any case (Hawkes, pers. comm., 1995).

Further references to Vavilov's activities

Professor A.D. Hall who succeeded W. Bateson as the Director the John Innes Horticultural Institute and had numerous meetings with Vavilov said, in 1930:

> Vavilov was a pupil of Bateson at the John Innes Horticultural Institution in 1913 and 1914... His special method was to use the geographical distribution of particular gene variations in determining the centre of their diversity, which he presumed to be the site of their origin. He applied this method to all the principal crop plants, leading expeditions to all parts of the world. This enabled him to deduce the centres of origin of these plants. The most interesting example of this method concerns the wheats. The cultivated forms were already known to fall into two inter-sterile groups, the tetraploids like *T. durum. T. turgidum*, *T. solonicum* and the hexaploids like *T. vulgare* and *T. compactum*. Vavilov found that the hexaploids had arisen from a Central Asiatic centre while the tetraploids had arisen from an entirely different centre in North Africa, hence the association of the two groups of wheat with different lines of human cultural development. The precise value of his methods and the importance of his "Law of Homologous Series in Variation", and of his principles of regional selection, are only beginning to be appreciated, but his published work undoubtedly represents the most important systematic attempt to study natural variation in plants since De Candolle and Darwin...
>
> Under his control this organisation (Vavilov's Institute) has come to embrace stations distributed all over the Soviet Union. Vavilov has used his centralised authority to secure a coordination of genetics, cytology, physiology and taxonomy, in which more than 2000 trained workers are engaged in problems of the efficient utilisation of plants in the national economy. The work of acclimatization and hybridisation carried out by the Institute bears witness to the success of his work, a success which is due to the combination of an enlightened method of enquiry with an industry, enthusiasm and organising capacity suited to the heavy responsibilities given him by the Soviet Government (Hall 1930).

Hardly anything can be added to the comprehensive character given to Vavilov by Professor Hermann J. Muller in his letter to M. Popovsky:

> All who knew Nikolai Ivanovich were buoyed up by his boundless good spirits, his generosity, his outgoing and sympathetic nature, his many-sided interests, and his energy. Colourful, lovable, and socially-directed, he transfused into others his zest for hard work, accomplishment, and cheerful cooperation. I have known no one else who created, managed, and developed ever further enterprises on such a great scale, yet who gave such faithful attention to the details. At the same time, Nikolai Ivanovich always adapted his methods of control to the human personalities involved, of which he was keenly aware. And most wonderful of all – he continued to keep up with the ever-growing branches of genetics and agriculture that his institutes and breeding stations dealt with, carried on copious reading and writing, organized varied researches, and collated sets of data...

> Nikolai Ivanovich once declared to me that "if you want to get something done by someone else, you should choose a man to do it who has already undertaken too much, for he is the one most likely to accept and accomplish that task too. One should," he said, "overload oneself also, for then one will get more done." This advice of his he certainly followed himself, and without sign of strain. Several times a year Nikolai Ivanovich returned to our Institute of Genetics in Moscow from one of his lengthy tours throughout the USSR, on which he had directed and checked up on the hundreds of plant breeding stations that were under his guidance...
>
> He was a truly great man, in very varied respects, scientific, administrative, human. Unlike some exceptional people, he was a thorough extrovert, without any perceptible trace of a feeling of inferiority or persecution or – in the attempt to compensate for them – of superiority. He lost himself in work, service, solving of problems, analysis and integration, perceptivity, and aesthetic appreciation. Having wide and deep awareness, he was also more life-loving, life-giving, and life-building than anyone else I have ever known. His effort and his example have not really been lost (Muller 1966).

Summarizing Vavilov's activities in a letter to the American Secretary of State Charles B. Davenport wrote:

> Vavilov has been regarded by geneticists everywhere as the leading geneticist of the USSR. His great learning, his broad ideas, his tremendous energy are of incalculable value not merely to the USSR but to agricultural science all over the world. Owing to the relation between the progress of genetics and that of national wealth in agriculture and other basic departments of national life, to interfere with the work of a man like Vavilov is committing not only national suicide but dealing a blow in the face to civilization (Davenport 1992, p. 514).

4 Vavilov's collecting missions and expeditions (1922–1940)

In his work Vavilov concentrated much effort and energy on the collecting and study of plant genetic resources. Expeditions to collect plant samples explored many regions in the USSR and abroad. He himself spoke several European and some Asian languages fluently, and was internationally acclaimed as an explorer. Vavilov personally led expeditions to more than 50 countries in Asia, Africa, Central and South America and collected a great number of seed samples of different crops. In 1938-1940 Vavilov intended to write essays on his journeys under the title *Five Continents*. A comprehensive plan of this book has been safely preserved, and to a considerable extent was carried out. It was, however, impossible to find a full manuscript. The existing pages were published in 1962 in Moscow (Vavilov 1987b), and have now been translated into English (Vavilov 1997).

The collecting missions by Vavilov and his researchers both in the USSR and abroad involved not only crop explorations, seed collecting, collecting site ecology and studies of cultivation technique, but also a geographical description of these countries and provinces and their various natural and meteorological conditions. Vavilov regarded the exploration of a certain territory as incomplete without ethnographical, historical and archaeological information, as well as without the data on the linguistic aspect of local names of crop varieties and species. The numerous samples delivered by collecting missions underwent testing and evaluation according to a unified pattern at the Institute's experimental stations located in different climatic zones of the USSR. Various data were summarized and published by the scientists of the Institute in the *Works*, or in monographs of individual authors. All of the work on collecting and studying plant materials, down to the smallest detail, was guided by Vavilov himself. The data resulting from the collecting of plant species and their complex study formed the basis of Vavilov's numerous papers and theories.

Vavilov founded his collecting efforts on geographical principles. In 1924, in his letter to G.K. Meister, Vavilov wrote:

> At present we have definitely accommodated ourselves to the geographical approach in crop research and the logically inevitable study of various areas, especially those in the countries bordering on Russia (Vavilov 1980, 164).

First of all, Vavilov was interested in the territories of the ancient agricultural civilizations and mountainous regions in various countries, where the mountain ranges and the regions of ancient agriculture were located.

After the expedition to Persia (now Iran) in 1916 and exploration of the adjacent regions, and Russia and Mongolia in 1922 and 1926, Vavilov understood that the most interesting materials were to be found somewhere to the southwest. This hypothesis was supported by the findings of V.E. Pisarev, a breeder from Siberia, whom Vavilov had brought into this work. Proceeding from the analysis of ample factual materials, Pisarev concluded that practically all Siberian varieties of cereal crops had originated in eastern Asia.

As a first step to proving this hypothesis, in 1922-1923 at Vavilov's initiative the Department of Applied Botany and Plant Breeding organized the first large-scale collecting expedition to Mongolia, led by V.E. Pisarev and V. Kuzmin. At that time, a civil war was on in Mongolia, and it made the collecting mission a complicated one. The expedition covered over 3000 km, explored all agricultural regions in Mongolia, and collected 1000 interesting plant samples. In the letters of that year Vavilov wrote:

> New centres (of diversity) for barley and millet have been found, as well as many endemic, hitherto unknown forms along with the ones the existence of which could have been predicted. For instance, wild millet possessing a similar hoof-shaped cross-section to that of wild oats, the previously unknown red-eared barleys, etc. were discovered (Vavilov 1980, p. 7l)

White-grained wheat and a special group of thin-husked and naked oats worthy of interest from both botanical and practical points of view have been found in northwestern Mongolia. It has been established that all agricultural regions of Mongolia show links with the mountainous provinces of Central China, and thus Siberian varieties can be traced back to China via Mongolia.

Despite his busy schedule Vavilov did not lay aside his collecting activities and explorations of Russia. In his letter to V.E. Pisarev dated 1922 Vavilov wrote:

> This year I managed to explore seven administrative districts, including Arkhangelsky, Pinezhesky, Kholmogorsky, Mezensky and Shenkursky (Vavilov 1980, p. 86).

In 1924 Vavilov thus launched a gigantic work on collecting cultivated plants; he planned to send collecting missions to Turkestan, a mission led by E.N. Sinskaya to the Altai, and by A.A. Grossgeim to Armenia to be continued by E.A. Stoletova in 1925-1926. In 1924, Sinskaya explored Altai, the plant resources of which had hardly ever been studied before. A multitude of leguminous, oil and fibre crop samples were collected there, 900 in total. Stoletova collected very interesting local cereal, leguminous, industrial, vegetable, forage and small grain varieties and forms in Armenia. In his letter to S.M. Bukasov, Vavilov remarked:

> Exceptionally valuable material (1700 samples) was brought from Armenia by E.D. Stoletova (Vavilov 1980, p. 245).

Afghanistan

The first among the most significant foreign collecting missions undertaken by Vavilov was undoubtedly the trip to Afghanistan organized in the second half of 1924. This country was especially attractive for Vavilov after his crop explorations in the Turkestan areas adjacent to Iran and Afghanistan performed in 1916. About one and a half years of difficulties had to be overcome before permission to enter Afghanistan was at last gained. Vavilov was scheduled to depart on the expedition in the second half of May 1923, so that he could reach the arid areas of Kandagar in Afghanistan in the period of cereal-ripening and harvesting. At this time he diligently studied the Persian language which was spoken by the chief officials in Afghanistan. While preparing for the mission to Afghanistan, Vavilov wrote a paper entitled *On the Eastern centres of origin of cultivated plants* (see Vavilov 1992) presenting a tentative review of the concepts of crop plant origin. Vavilov failed to enter Afghanistan in 1923 because of political reasons; the relations between the USSR and Afghanistan had become strained. At the end of 1923 Vavilov planned that, if they failed to enter Afghanistan in 1924, they would explore the bordering area and make a transverse section across the most interesting new areas (from the point of view of exploration). During the exploration of Turkestan, Vavilov planned to try to enter Afghanistan through the border posts. In a letter to P.M. Zhukovsky in 1924 he wrote of his plans for the future after the failure of his trip:

> If in this year you make a survey of Asia Minor while we explore Turkestan Region, it would be possible to find many interesting things. The success of this year would determine the opportunity to organize collecting in China and Africa (Vavilov 1980, p. 161).

The permission and funds for the collecting mission were allotted in early 1924. However, difficulties in obtaining the entry visa to Afghanistan delayed departure. Applications had to be made to the Embassy of Afghanistan several times, but the Embassy did not dare to take responsibility for issuing visas for the collecting team under the pretext that if the mission was granted permission it would lead to inquiries from England and other countries. Only thanks to the exceptional assistance rendered to the collecting mission by the People's Commissariat of Foreign Affairs, were the team members listed as Plenipotentiary Diplomatic Representatives of the USSR in Afghanistan, and V.N. Lebedev, a breeder of the Variety and Seed Division of the Sugar Trust, and D.D. Bukinich, an agronomic engineer, listed as couriers of the Commisariat of Foreign Affairs.

At long last, on 19 July 1924 the collecting team passed across the border of Afghanistan in the vicinity of Kushka. In August and September 1924, military activities started in Afghanistan. Warfare spread over the whole southern part of the country while half of the country was terrorized by the Basmatchs. To make collecting worthwhile it was necessary to have guards whose wages had to be paid at the mission's expense. Thus, the team was constantly accompanied by two or three Afghan soldiers, and in some places by even ten guards. Only under such terms did the mission manage to accomplish its tasks to the fullest extent. In the same year, on the first day of December, the mission returned to Kushka after completing explorations along the scheduled route (Vavilov 1987b).

Explorations covered all the territory of Afghanistan. Partly as a whole group and partly after splitting into smaller groups, the collecting mission managed to travel about 5000 km. This mission visited all the regions typical of Afghanistan, and also explored Kafiristan in the southeast of the country, which was practically unknown to Europeans at that time. From the extreme verge of crop distribution at the altitude of 3500 m, the team travelled down to the subtropical and even tropical areas bordering on India.

For the first time, agricultural exploration could cover the vast areas which stretched for 400 leagues along the border with Turkestan which had been previously completely inaccessible for Russian explores. Beyond the Hindu Kush, near India, they discovered a large area with totally unique wheat and rye varieties unknown in Europe. They collected numerous samples of leguminous, oil-bearing and horticultural plants. Afghanistan also proved to house the richest diversity of bread wheat (*Triticum aestivum* L.) in the world. Many new varieties of *T. compactum* L. were also found there. The results of this collecting mission are described in the book entitled *Agricultural Afghanistan* (Vavilov 1959) and in the sketches included in *Five Continents* (Vavilov 1987b; 1997).

In his work *Agricultural Afghanistan* Vavilov wrote:

> In South-Eastern Afghanistan and adjacent regions of India there is the main centre of diversity of bread wheat varieties – the major baking crop of the Earth. The true predecessors of cultivated rye were also found here (Vavilov 1959, p. 50).

Fig. 25. Collecting mission in Afghanistan (1924). Kabul. N.I. Vavilov is sitting among the officers of the Soviet Embassy on the back seat without a hat [VIR Archives]

As for other cereals, Vavilov did not find genuinely distinct barley forms, and cultivated oat was completely absent. However, the regions of Afghanistan beyond the Hindu Kush were of great interest in terms of the diversity of local varieties and many other groups of plants. Even initial perfunctory surveying led to the discovery of a number of new varieties of horticultural, leguminous, oil-bearing, medicinal and industrial fruit-bearing plants as well as some arboreal species. Many cultivated plants revealed all phases of their evolution. And in the foothills of southwestern Asia, a striking diversity of forms was discovered in grain legumes, namely in horse beans, pea, lentil and cowpea. Everything pointed to the existence here of the focus of form development and origin of these crops. High in the mountains were found forms showing similarity in height and branching habit with intermediate forms of cultivated flax. There were unique forms of Afghan wild cabbage, rocket (*Eruca* spp.) and cress, and the transitional forms from weed to cultivated plant. The diversity of melon with its transitional forms showed the way from wild melons to cultivated ones and helped to identify the primary area of form development for this crop. Similar phenomena were manifested in many other crops. Major areas of diversity for such cultivated plants as wheat, pea, beans and other field and garden crops, were located mainly at altitudes between 1000 and 2000 m.

The richest diversity of cultivated forms untouched by selection, with roughly expressed semi-wild traits, occurred in southeast Afghanistan, the Punjab, Kashmir and adjacent northwest India. Here, in the geological fold between the Hindu Kush and the Himalayas, with its striking diversity of environments, abundance of crop genes and variety of ethnic groups, Vavilov identified a primary focus of agricultural civilization. Preliminary results of the expedition to Afghanistan and subsequent detailed analysis of all crops collected fully proved assumptions about the important role of ancient foci located high in the mountains and in the foothills of southwest Asia.

After returning from the expedition, Vavilov made a number of public presentations, for example in the Russian Geographic Society which awarded him the N.M. Przhevalsky Medal "For a Geographic Feat" recognizing his extremely hard trip to Afghanistan. However, to make the knowledge gained more comprehensive it was necessary to extend explorations, if possible, to the oases of Central Asia. In 1925, in a letter to P.M. Zhukovsky, Vavilov wrote:

> On 1st of July or so I'll be in Turkestan where the Afghan materials are planted near Tashkent, and intend to find my way to Khiva, Fergana, and some places where I haven't been before, and ... for a short time in Western China with a small company whom I'm to assemble on my way to Kashgar and Khotan and who would further proceed on their own to Urumchi and Kulja. The opening of the Consulate and invitation from the Foreign Affairs Commissariat to explore Western China have unexpectedly brought this task, as a part of the programme of explorations in the whole of China in general, before me (Vavilov 1980, p. 191).

Thus, in 1925, Vavilov accompanied the agronomist V.K. Kobelev to the oasis of Khoresm. This area attracted him by its isolation and traces of an ancient culture. After very detailed examinination from 1926 to 1928 of the plantings of over 1500 samples collected, Vavilov published the results of his explorations in Khoresm in the article entitled *Cultivated Plants of the Khiva Oasis: Botanical and Geographical Essay*. Analysis of the composition of the cultivated plants in Khoresm allowed Vavilov to conclude that they did not represent independent crops. Comparing the cultivated plants of Khoresm with those of Iran, Afghanistan, Tadjikistan, Uzbekistan and Turkmenistan gave evidence of kinship. Thus, the origins of the cultivated plants in Khoresm were to be sought in these regions.

Transcaucasus and Turkey

In addition to Afghanistan Vavilov became deeply interested in cultivated plants of the countries adjacent to Russia, namely, Asia Minor. Beginning in 1922, Vavilov planned an exploration of the Transcaucasus and Asia Minor. In this year he wrote to P.M. Zhukovsky:

> It is extremely important that you should also explore the regions closest to the Transcaucasus, Turkish Armenia, and the whole of Asia Minor... (Vavilov 1980, p. 89).

In 1924 Vavilov started raising funds for a collecting mission to Asia Minor. In a letter of April 1924 addressed to the Variety and Seed Division of the Sugar Trust, Vavilov explained the need for a collecting mission to Asia Minor:

> Recent investigations have clearly shown that agriculture in Europe and America is based quite often on agricultural plant varieties having little value and, besides, not used by the breeders on a sufficiently wide scale. On the other hand, these investigations have revealed in nature a gigantic store of varieties, as yet unused in the world agricultural industry. For instance, numerous forms of field plants which the science and practice of the developed countries have been unaware of, were discovered in the regions of south-western Asia, western Asia and the Transcaucasus. Cereal grasses of Asia and the Transcaucasus are of great interest from a practical point of view. They are characterized by non-shattering, drought resistance, excellent vitreous grain, tolerance to soil quality, and immunity to many parasitic fungi...

> ...The Department approaches the Seed and Variety Division with a proposal to organize this summer (1924) a collecting mission to the regions of Asia Minor and South Transcaucasus with the aim of collecting and exporting seed materials of all field crops. As leader of this collecting mission, the Department of Applied Botany recommends a learned specialist Prof. P.M. Zhukovsky, the former director of the Tiflis Botanical Gardens and expert in the cultivated flora of the Transcaucasus; he has been engaged in investigations in this region for a long time (Vavilov 1980, p. 159-160).

Zhukovsky's expedition to Turkey took place in 1925 and continued in 1926 and 1927. It was one of the first collecting missions aiming to study and collect both cultivated and wild plant species in countries with ancient agriculture. The first year's explorations covered the whole western part of Asia Minor including Cilicia and the Ankara region. The following years were devoted to exploring northern Mesopotamia, Syria and the island of Rhodes. Zhukovsky collected over 10 000 samples along a 12 000 km route. As a result, the world collection was replenished with valuable forms of bread and durum wheats, protein-rich forms of barley, red oats, table and canning peas, chickpea, lentils, anthracnose-resistant *Phaseolus*, ultra early cowpea, high-yielding vetch, tender-stemmed alfalfa, candlestick-shaped winter flax, oil-rich sesame, mustard, and winter rape, morphine-rich opium poppy, sugar-rich cassava and cantaloupe melons. For the correct identification of the materials collected, in 1928-1929 Zhukovsky undertook a study of herbaria and literary sources at the Berlin Botanical gardens and in the natural history museums of Vienna and Paris.

The results of investigating the plant resources of Asia Minor were reflected in a series of papers by Zhukovsky and in his fundamental monograph *Agricultural Anatolia* (Zhukovsky 1934), written by Zhukovsky himself and by his co-authors who studied the collected materials. There is scarcely a monograph in the literature devoted to Asia Minor that can be compared with this one.

North Africa

From as early as his visit to Great Britain and work with Percival's collections from Abyssinia, Vavilov felt the need to travel to Africa. In 1922, in a letter to Yu. N. Voronov he wrote:

> I have a personal interest in southern crops and intend to organize in 1925 a collecting mission of the Department to North Africa. I have tried to procure as much literature about the southern countries as I can. We are keeping Africa in mind mostly for examining field crop plants (Vavilov 1980, p. 84).

Vavilov gave himself the great task of visiting all of the Mediterranean countries and those of North Africa. There he hoped to collect as diverse as possible materials of the local varieties, and to study their history and cultivation conditions. However, it was not easy to make such a plan a reality. Most of the African countries were colonies and entry was denied unless sanctioned by England or France. Usual diplomatic contacts were of no avail; Vavilov had to obtain entry visas by himself.

In the early summer of 1926, as soon as he received permission from the Soviet Government and obtained funds needed for the journey, however limited, Vavilov departed for London where he managed to get visas for Palestine and Cyprus with the help of his friends C. Darlington and especially A.D. Hall, head of the John Innes Institute. In a letter of 1926 Vavilov wrote the following:

> ...the visa issuing will be solved favourably for us, because before his death Bateson requested Dr Hall, the research leader at the Ministry of Agriculture, to make all efforts to help us (Vavilov 1980, p. 257).

In France, Vavilov's application to the Ministry of Foreign Affairs proved fruitless. However, invaluable aid was provided by Academician A. Chevalier and Mme. F. De Vilmorin, the head of the famous seed company Vilmorin et Andrier. She herself visited President Poincaré and Prime Minister Briand and took pains to get visas for Vavilov to Algeria, Morocco, Tunisia and Syria.

By mid-June 1926, Vavilov was already on his way from Marseille to Algeria. Narrating his trips to Algeria in his book *Five Continents*, Vavilov wrote that it was the sultriest season when hardly anybody dared to travel over the countryside. As for himself, not withstanding the dissuasion of L. Trabut and L.O. Ducellier, he set out on his journey. When travelling through Algeria, Vavilov came to a conclusion on the need to separate the mountainous, foothill and maritime territories in North Africa. Thus, it should be taken into consideration that mountainous areas were likely to possess more ancient agricultural systems than maritime ones. In the coastal fields Vavilov found enormous bulbs of common onion which reached 2 kg. Beans, lentil, cowpea, wheat, barley, flax, wild carrot and weedy vetch were notable for their unusually large flowers, seeds and fruits. Large size was a typical feature of cultivated Mediterranean oats and wild weedy oats. In Vavilov's opinion it was the result of natural selection and the agricultural practices of the population. In the regions of Setif, Timgad and Tiaret, a significant area of durum wheat was collected by Vavilov with the help of Ducellier and Trabut.

From Algeria, Vavilov moved to Morocco where he was planning to survey its major agricultural region in 10-12 days. On his way to the Atlas Mountains, Vavilov discovered a peculiar form of durum wheat with easily shattering seeds and also found rye, hemp, pea and cowpea. All evidence attested to an undoubted linkage of agriculture in the mountainous areas of Africa not only with the great Mediterranean culture, but also with that of southwest Asia.

In Tunisia, Vavilov received from Professor F. Boeuf, Director of the Botanical Gardens, diverse and numerous samples of cultivated plants. Together they travelled all over the country. The season was extremely favourable for such a journey: harvesting in the mountains had just begun and cereals were ripe in the field. Vast spaces of the Tunisian foothills and uplands appeared to be planted with durum wheat. There were only old local landraces under cultivation with variegated mixtures of numerous subvarieties. As in Algeria and Morocco, Tunisian agriculture proved to be rather primitive. However, northern Africa as a whole represented a certain entity. Botanical and geographical analysis demonstrated the uniqueness of the Mediterranean culture and revealed the domination of original local durum wheat along with large grain and six-row barley. Cultivated coarse-grained legumes and coarse-seeded flax were concentrated in the coastal area. The regions of the Atlas mountains showed the twofold influence of the southwest Asian and Mediterranean cultures. Comparative uniformity of cultivated plants and the extensive influence of local agriculture were witness to the fact that cultivation had not started there. The source of cereals should be sought in western Asia.

The Middle East

After his explorations in North Africa Vavilov returned to Marseilles. With a French entry visa to Syria he departed by sea to Beirut. The itinerary in Syria had been worked out by Vavilov in London and Paris. It was important to reach the inner parts of the country: the southern regions of Syria, the zone bordering Palestine, the mountains where in 1906 Dr A. Aaronsohn, a botanist, had discovered wild wheat for the first time; the Khoran uplands, which were reported to be one of the major areas of cereal cultivation and, at the same time, of the origin of wild wheat. It was the place where the first collecting of a particular subspecies of durum wheat, later known as Khoran wheat, had taken place, and the dense populations of wild wheat (*Triticum dicoccoides* Koern) had been found. Bouts of malaria were a severe disturbance to Vavilov and prevented him from continuing to collect shattering wild wheat and wild barley (*Hordeum spontaneum* C.K.). In spite of his illness, however, Vavilov collected all the seeds of interest to him and proceeded to Damascus. Because of the siege of Damascus at that time (begun after the insurrection of the Druse), Vavilov had to confine himself to examining the downtown seed markets and visiting only a few fields. The range of collected wheat varieties was extremely diverse and showed the influence of both the southwest Asian and Mediterranean civilizations. He discovered very peculiar Mediterranean mountain forms of cowpea which had been used there instead of fodder barley. Having collected and mailed numerous samples, Vavilov moved to northern Syria over the lands of Mesopotamia, with the purpose of crossing the valley of the river Euphrates by car.

All of this region was the granary of Syria, conventional Mediterranean agricultural practice prevailing, with mostly wheat and two-row barley under cultivation. The flora was quite different from that of the southwest Asia and the Iranian/Turkestanian region. In Latakia and the Lebanon Mountains he found curious wild oats, perennial rye, wild peas, wild olives and carob. While in Damascus, in a letter to V.E. Pisarev of 23 September 1926, Vavilov shared the misfortunes that had happened during this journey:

> Please be informed that to my deep regret I have caught malaria somewhere in Crete or Cyprus. I feel rather unwell. Now I am hurrying to Beirut where they will start injections. Syria began to give way. The military authorities permitted me to enter the battlefront area at the border with Palestine. There I've found *Tritcum dicoccoides* Korn. It is obviously widely distributed in Syria and Palestine (Vavilov 1980, p. 287).

Waiting for a visa for Egypt and Abyssinia (Ethiopia) forced Vavilov to linger in Palestine and Transjordan for two months. When the planned trip to Jerusalem and Tel Aviv was finalized, Vavilov and Dr Eytingen set out to the valley of Ezdralion. The Palestinian wild wheat was absolutely different from the Khoran type: ears and spikelets were larger just as in cultivated wheat, but the awns were coarse and grains large.

> The very fact that wild barley occurs with wild wheat is witness to the fact that Palestine, like Syria, really belongs to the basic area of origin of the world's most important cereals: wheat and barley (Vavilov 1987b, p. 93).

Vavilov wrote about his attempts to get an Egyptian visa in his book *Five Continents*.

> Next in turn was the trip to Egypt. Endless attempts to get an entry visa there were a failure. Mr Mossari, the banker, in spite of all his influence was unable to gain the required permission for us. The aid of Mr Kourdali, President of the Arab Academy of Sciences in Damascus, was also in vain. Petitions of the most prominent English

> agronomists, Daniel Hall and John Russell, were of no help either. I have received quite courteous replies from Alexandria signed by the British colonel who was in charge of entry permissions for foreigners that unfortunately the existing circumstances at present thwarted the possibility of entering the country (Vavilov 1987b, p. 107).

Vavilov had to content himself with sending an Italian student, Mr Gudzoni to Egypt in order to collect seed and fruit. This man proved to be an invaluable help to Vavilov collecting and shipping numerous seed collections from Egypt to the Soviet Union and exploring the whole route outlined by Vavilov, including all agricultural regions of Egypt up to Aswan.

Abyssinia and Eritrea

Vavilov now made all possible efforts to obtain a visa to Abyssinia and Eritrea and wrote:

> Preliminary discussions in Paris were fruitless. Mme. De Vilmorin promised to send a letter to the French Ambassador in Addis Ababa and, as I have later realized, fulfilled her promise with the kindness so typical of her. The difficulties were aggravated by the fact that at that time Abyssinia had no diplomatic missions in Europe. The endeavours to send cables and letters to the Abyssinian Government from different countries also failed. Our friend Dr Harlan, an American agronomist who visited Abyssinia in 1923 and was heartily welcomed by the Ethiopian ruler, tried everything on his part in Washington to help us, but evidently it was one more voice crying in the wilderness of Abyssinian diplomatic circles. It was probably reasonable to reject the utopian idea of visiting Abyssina, but I could not reconcile myself to this, as all our fundamental postulates attested that eastern Africa must be characterized by a unique cultivated plant diversity which was as yet completely unexplored but known only by random excerpts from floristic research works. The International Agricultural Institute in Rome to which I was advised to apply, helped me to get a visa to Eritrea, an Italian colony, but stated that affairs with Abyssinia were beyond their scope. Nevertheless, this was something really palpable and aroused further expectations (Vavilov 1987b, p. 107).

The way to Abyssinia was through French Somalia crossing by the railroad from the Red Sea port of Djibouti to Addis Ababa, the capital of Abyssinia. However, in order to take a steamship to Djibouti with a French visa, Vavilov had to return once more from Palestine to Marseilles and cross the Mediterranean Sea. As noted before, all his life Vavilov was a poor sailor and every time suffered from seasickness. This is more evidence of Vavilov's persistence in achieving his goals.

On 27 December 1926, Vavilov set out from Djibouti by railway to Addis Ababa without a visa to Abyssinia. However, he was to interrupt his journey and moved by foot through Harar to the capital of Abyssinia. Everything appeared absolutely atypical there: the cultivated crop diversity, wild plants, agricultural techniques and climate. Wheat was represented by uncommon varieties and even by uncommon species. Wheat and barley fields demonstrated an exceptional mixture of forms. Vavilov found there an original endemic grass – tef (*Eragrostis* Trott), which was used to produce excellent flour for pancakes; an oil-bearing plant named noog (*Guizotia abyssinica* Cass.) with black seeds, and so on. Distinct diversity of sorghum was also observed there.

After a one-week trip through Harar, Vavilov dispatched to Leningrad his first 40 boxes, each containing 5 kg of seeds of various cereals and other plants. Upon arrival at Addis Ababa the sanction of the national leader Regent Ras Tafari had to be secured. With the assistance of the French Ambassador who had received Mme. De Vilmorin's letter from Paris the problem was settled. Vavilov obtained the so-called "open folio" which proclaimed him a guest of Ethiopia and granted him leave to travel over the whole country. Vavilov himself was received by Ras Tafari and had a long and interesting talk with him. Vavilov described this conversation as follows:

> With great interest did Ras Tafari question me about my country. His particular attention was drawn by the Revolution and the fate of the Emperor's court. We told him all the story known to us in brief. It's difficult to imagine a more attentive listener. As to a most engaging fairy-tale did the Ethiopian ruler listen to the brief story about our country and the events that had happened there (Vavilov 1987b, p. 111).

While in Addis Ababa, Vavilov organized small-scale explorations in the vicinity of the capital with the assistance of carefully selected and specially instructed people who collected quite a lot of seed samples of cultivated plants for him. Thus it became possible to acquire plant materials from the most inaccessible places. As a result, numerous interesting forms and varieties were discovered. Seeds were also acquired at the city market. Preparations for the journey through Abyssinia were at last finalized and on 7 February 1927 Vavilov led the team of 11 members armed with rifles and light spears and 12 mules from Addis Ababa toward Ankoberi. The time chosen for the trip proved favourable not only in terms of the ripening of cereals, but also in the seasonal distribution of the population. Vavilov wrote:

> Interior Abyssinia such as the Gonder region is full of endemic species. There are enormous crops of Abyssinian teff, interesting, peculiar, and variable wheat in an unbelievable mixture of forms, and mixed crops of barley, including black naked-seeded ones not known anywhere else in the world except in this country. There are large quantities of peculiar local Abyssinian forms of lentils, chickpeas, peas, and vetches. Large bushes of wild castor bean plants (*Ricinus communis* L.) usually grow along fences. There are also peculiar cabbages and wild mustard (*Brassica carinata* A. Braun.) producing a large quantity of seeds but at the same time used for both the roots and the leaves. Spelt (*Triticum spelta* L.) is also abundant (Vavilov 1987b, p. 113).
>
> In contrast to Abyssinia, there is a great variation of climates, soils, and other natural conditions in Eritrea... The composition of bread grains and leguminous plants is quite similar to that typical of Abyssinia, but all present some differences. Mountainous Eritrea is similar to Abyssinia with respect to the variety of plants cultivated, but here the influence of Europe is telling. Northern Abyssinia and the mountains of Eritrea were occupied by the Portuguese during the fifteenth century... Actually, the Portuguese introduced the cultivation of peppers here (*Capsicum* spp.), which have become naturalized both in Eritrea and in Abyssinia (Vavilov 1987b, 119).

Upon completing his four-month trip in Abyssinia and Eritrea which lasted from December 1926 until April 1927, Vavilov summarized his results as follows:

> There is no doubt that this comparatively small mountainous territory represents an independent centre of agricultural civilization. Although contemporary historians and

archaeologists are inclined to view the Abyssinian cultures adopted as secondary, a study of the specific and varietal composition of the cultivated plants and the agrotechnology shows the opposite. The presence of native endemic species such as teff, chickpeas, Abyssinian bananas [*Ensete ventricosum* (Welw.) Cheesm.], and a species of mustard (*Brassica carinata* A.Braun.), as well as an absolutely original species of wheat differing both genetically and anatomically as well as by its complex of characteristics all lead necessarily and logically, on the basis of comparative studies, to the acknowledgment of a mountainous Abyssinian centre, a deservedly independent division...

A number of crops (teff, chickpeas and bananas) arose no doubt just in the Abyssinian centre; however, there is no wild wheat or any wild barley and no wild beans, so that, possibly, the origin of these plants is linked to other, adjacent territories, primarily within the Near East in a wide sense. However, there is no doubt that the isolation of the cultivated plants within Abyssinia took place in the distant past. This is indicated by the presence of such endemic types as violet-grained wheat, the multitude of endemlc characteristics distinguishing the Abyssinian barley, and such anatomical characteristics as a low number of vascular bundles in the coleoptiles (Vavilov 1987b, p.199)

When comparing the foci of land cultivation in the Old World, Vavilov acknowledged the necessity of identifying Abyssinia and the adjacent mountainous Eritrea as an autonomous centre. The absence of a large diversity of Old World fruits and vegetables proved the uniqueness of the Abyssinian agricultural tradition. Direct explorations revealed the exceptional value of the Abyssinian barleys resistant to European infectious diseases and characterized by resistance to lodging, large size of grain and indifference to warm temperature. Great attention was paid to Abyssinian peas, especially fodder ones which yielded large amounts of green matter and were ready for green manuring. Extremely interesting were awnless forms of durum wheat. Vavilov noted that the evolution of Abyssinian vegetation and that of mountainous Eritrea were undoubtedly somewhat united. According to Vavilov's judgement, the plant genera of Cape Province, Abyssinia and mountainous Eritrea had many elements in common with the Himalayas and the Mediterranean. It was obvious that this region and the adjacent areas required more comprehensive exploration. It was necessary to trace the links of these territories with Yemen where vegetation evidently blended elements of both southwest Asia and Abyssinia.

Having mailed 80 5-kg parcels with seeds and ears from Asmara, Vavilov set out for Massawa on the Red Sea Coast.

The Mediterranean

After exploration of the African continent; Vavilov faced the task of studying the agriculture of the Balkan, Apennine, and Iberian peninsulas, the three major ones in southern Europe, as well as of the largest islands in the Mediterranean Sea. Most of Greece was occupied by vineyards and olive orchards. The seed market in Athens offered a mixture of mostly local Mediterranean crops with varieties borrowed from western Europe and America. When travelling through the Thessalian valley, Vavilov observed boundless fields of wheat, barley, and grain legumes. The varietal and crop composition tended to change with ascent into the mountains. Down on the plain there was the kingdom of bread wheat, while higher up durum wheat prevailed. In the foothills, with the increase of rainfall, cone wheat (*Triticum turgidum* L.) dominated.

The typical Mediterranean set of cultivated plants was disappearing. Signs of transition to Macedonia and to the steppes of southern Europe could be traced. In any case, Greece was never a big agricultural centre; its culture was based mostly on tree species, namely olives, carob (*Ceratonia siliqua* L.) and grapes. From Athens, Vavilov proceeded to Crete where the greater part of the cultivated land was occupied by cereal crops. Mostly grapes and carob were grown in the mountainous Southern part. A number of endemic forms of grain legumes, especially of cowpea were found on the island. The Mediterranean large-flowered, large-seeded flax was cultivated over vast areas.

Vavilov described his findings as follows:

> At high altitudes interesting endemic 'forms of nonligulate hard wheat have developed. The position of the island favours the distinction of special simplifications, so-called recessive forms of great variety. As we have seen in the case of soft wheat, the centre of formation of nonligulate wheat is found in the Pamirs, an isolated part of Inner Asia. In the case of Mediterranean hard wheat, the islands have assumed the role of isolated areas (Vavilov 1987b, p. 124).

From Cyprus Vavilov proceeded to Italy which he had visited many times before. He explored Sicily along the route from Palermo to Catania and carried out a thorough exploration of Sardinia. Vavilov was convinced that the exploration of Italy and its islands had decisive importance in order to understand the development of Mediterranean culture.

Fig. 26. N.I. Vavilov. Mediterranean collecting mission, Greece, 1926 [VIR Archives].

Spain

After Italy, Vavilov went to Spain. He sailed from Genoa to Barcelona on 27 June 1927. He received a most cordial welcome there from the scientific community, especially by Professor P. Bolivaz and his son, as well as by Professor Crespi, a botanist who accompanied Vavilov on some of his journeys in Spain. As on his visits to other places, Vavilov pursued the major task of familiarizing himself with all agricultural regions in Spain. To this end, he had to cross the country in all directions and collect the maximum possible seed materials of field, vegetable, and other crops.

In his book *Five Continents*, Vavilov recalls an interesting episode. In those times, Soviet visitors were treated with a good deal of suspicion, and Vavilov mentioned this several times in his notes. In this respect, he recalled:

> ...he (Prof. Crespi) approached me with an embarrassed look and declared that he wished to talk to me about a secret matter. It turned out that the agents who had been following me all the way from the Spanish border were convinced of my peaceful intentions and had asked Professor Crespi to talk to me about concluding an agreement. These persons had declared that the Russian professor was exhausting them by his rapid movements by automobile and train and on horseback in the mountains, because of which, uneasy about their health, they wanted to suggest the following compromise: the professor should let them know in advance where he was headed and what his destinations were so that, although officially they were supposed to accompany him, they would not need to follow him in the mountains and in particular when on horseback, but could wait for him at specific places at inns or in cities. In return they promised to assist in every way possible with travelling purchasing tickets, reserving hotel rooms, and sending off parcels. After deliberating about the suggestion, I decided to conclude the deal. I was introduced to them. I had already noticed these familiar characters in bowler hats and official garb. The first days after the agreement was concluded went comparatively well. I kept myself occupied mainly in the mountains, while they spent their time with great pleasure in cities and hotels........Later the agreement was broken owing to their constant habit of ordering rooms mainly in very expensive hotels in the centre of cities and to their general tendency to live the high life (Vavilov 1987b, p. 128-129).

In Madrid, Vavilov got acquainted with the famous Madrid Botanical Gardens, and in particular, with its unique herbariurn collected by La Gasca as far back as 1818:

> The herbarium of La Gasca is the best of the old herbaria of cultivated grasses; with its help it is to a great extent possible to reconstruct the composition of the vegetation of Spain at the beginning of the nineteenth century. At that time Spain was ahead of all other countries with respect to knowledge of cultivated plants (Vavilov 1987b, p. 199).

Further, Vavilov wrote:

> I cannot forget the noble action of the families of La Gasca and Cavanilles, to whom 1 turned with a request for help in acquiring a rare book published by the families of the most famous botanists of Spain. In response to my request I received a touching letter: it stated that the families owned only a single copy of this book but, after considering my request, they had decided that a book like this was more necessary for a botanist and that they were giving it to the Russian professor with their best wishes for the flowering of Soviet sciences (Vavilov 1987b, p. 129).

Madrid was chosen as the base from which Vavilov made lengthy trips to different regions of central Spain where cereal crops were widely cultivated, and thus he visited almost all Spanish provinces and made a short visit to Portugal. He started with the southeastern part of the country and finished with the northern regions, i.e. Galicia, Asturias and Basque Provinces.

Galicia is the rainiest Spanish province with rich arboreal and meadow vegetation. Field crops differed greatly from those in other parts of the country. The diploid small oat of Galicia (*Avena brevis* Roth., *A. strigosa* Schreb.) was described by Vavilov as a "characteristic world endemic". Genesis of this crop here and in northwestern Portugal was linked by the scientist to wild oats close to the latter in terms of origin. Thick-growing wild flax, genetically most close to the cultivated one, as well as perennial kale were encountered in Galicia. In contrast to southern and inner Spain, varieties of cowpea, lentil, chickpea and pea were found to be of an apparent Asiatic origin, probably brought from either Transcaucasia, or southwestern Asia in very distant times. They differed strikingly from the original large-seeded forms of southern Spain. Unlike in central and southern Spain, fields of flax for fibre were sometimes encountered. Potatoes, barley and maize were grown in abundance.

In Asturias, the northernmost part of the country adjacent to Galicia, the culture of true emmer (*Triticum dicoccum* Schuebl.) has survived. Asturian emmer differed from that in Tirol and Bavaria, as the former was not a winter emmer but a spring and prevailingly awned one. Vavilov recalled:

> I happened to arrive in Asturias just at the time of the emmer harvest. To my surprise it turned out that this crop was harvested not by means of sickles or scythes, but by using ancient flails by means of which the spikes are removed and then thrown into baskets. During all my travels through the years in some sixty countries not once had I been fortunate enough to observe such a mode of harvesting. It was only later that I met with a similar method in the mountains of western Georgia, in the village of Lechkhum, where an important endemic group of wheat, including a special species genetically very close to the present emmer, was recently discovered (Vavilov 1987b, p. 142).

Asturias, with its historical stages in the evolution of agriculture, was seen by Vavilov as a unique place in Europe, undoubtedly worthy of special attention by researchers. From Lugo (in Galicia) Vavilov drove to Pamplona, the country of the Basques. The composition of field crops here revealed many original things. It was the kingdom of emmer wheats and peculiar oats encountered nowhere else. There was a sharp difference between the Basque provinces and eastern Spain. Wheats manifested the widest diversity, and cone wheats (*Triticum turgidum* L.) were observed quite frequently. Forage crops, like alfalfa and red clover, were cultivated in abundance. Near Pamplona, Vavilov discovered the results of spontaneous mass hybridization of bread wheat with an *Aegilops* species. True emmer was not sown in the Basque Provinces.

Upon completing agronomic and botanical investigations in Spain and Portugal, Vavilov came to the following conclusions:

> Spain turned out to be an exceptionally interesting country for understanding the evolution of European agriculture. There I was able to establish without doubt the presence of a number of endemic crops, described from the Iberian Peninsula alone: sand oats (*Avena strigosa* Schreb), special species of vetches (*Vicia articulata* Hornem. and *V. narbonensis* L.), einkorn (*Triticum monococcum.* L.), and fodder plants such as gorse (*Ulex europaeus* L.) and chesnut (*Castanea sativa* Mill). In addition, during the stages of their development some crops contained weedy plants, closer to other, more ancient crops. This was particularly clear in the case of oats. In Spain it is possible to trace various stages of

agriculture up to the present, starting with primitive field work, harvesting, and milling. As demonstrated by a comparative study of the Near East and other countries, the overwhelming majority of the basic crops were introduced into Spain. This occurred thousands of years ago. The influence of Roman, Syrian, Egyptian, and Arabian crops can be successfully traced. Spain, so to speak, absorbed all of Mediterranean agriculture and, in part, reworked it, creating its own new varieties. The fact that the material here appears basically to be introduced is indicated by characters and by the lack of any complete taxonomic systems of species (Vavilov 1987b, p. l44-145).

Vavilov was of the opinion that the intensive agriculture of Eastern and Southern Spain undoubtedly provided for the breeding of wonderful varieties of the large-sized onion of Valencia, large-seeded grain legumes, chickpea, beans and chickling vetch in particular, as well as olives, to which more attention should be paid by the breeders.The antiquity of the country and diversity of crops determined the richness in varieties. A study of the range of cultivated plants in Spain and a comparison with varieties from other countries of Europe, Asia and Africa clearly demonstrated the influence of migrations and borrowings on this range. At the same time, Vavilov also noticed the presence of an independent culture. In this respect, the Iberian peninsula appeared to be one of the most interesting places in Europe.

In mid-August 1927, Vavilov left Pamplona for France via the port of San Sebastian, and then proceeded to Germany. There, together with Professor Baur, Vavilov made a short tour of the mountainous regions of Wurttemberg in order to clarify some questions related to the origin of *Triticum spelta* L. Finally, he returned to Russia in late August.

In 1926-1928, V.V. Markovich, an expert in tropical fruits, explored Palestine and the areas of Punjab, Kashmir and Assam, following Vavilov's theoretical guidance. He also managed to visit many other states of India, and the islands of Java and Ceylon. As a result of that undertaking, the Institute received approximately 3500 samples of mainly cultivated and wild tropical plants and also of vegetable and fruit crops (Vavilov 1980).

The Far East

At Vavilov's initiative, in the autumn of l928 and the spring of 1929 E.N. Sinskaya of the Department of Applied Botany made two trips to Japan. Those brief expeditions enabled her to obtain many samples of cruciferous, textile fibre and oil-bearing plants and root crops as well as to greatly replenish the collections of wheat, barley, oat, maize, rice, sorghum, soyabean, hemp, ramie (*Boehmerta nivea)*, watermelon, melon and pumpkin with over 2000 accessions. These materials included a series of new forms exceptionally valuable for the USSR. The Japanese tangerines brought by Sinskaya were used for breeding cold-tolerant varieties at the Institute's station in Sukhumi. The collection of persimmon gathered in Japan was planted at the station in Zakataly (Azerbaijan), while the pear scions were used for inoculations at the Maikop Experiment Station of VIR (Soratniki 1994).

After his long trip over the Mediterranean and East African countries Vavilov had to take a break to put the collected scientific materials in order and resolve a number of research and organizational problems which had arisen in his absence. However, by 1929, Vavilov together with botanist M.G. Popov undertook his next collecting trip to western China (Xinjiang) and then went alone to Japan, Taiwan (the island of Formosa) and Korea. During his explorations in western China he also surveyed some regions in Kirghizia and Kazakhstan. The purpose of the mission was to find out about the cultivated plants of western China and other oriental countries, and make collections as representative and comprehensive as possible.

In July 1929, Vavilov's caravan equipped in southern Kirghizia moved along the Alai Vale toward the frontier checkpoint of Turkestan and further on to the first scheduled goal of the journey, the town of Kashgar in Kashgar Oasis, situated at an altitude of 1200 m. As a result of explorating Kirghizia, he wrote a paper entitled *Plant industry in Soviet Kirghizia and prospects for its development*. For instance, he wrote:

> Upon crossing a difficult mountain pass, a traveller enters Soviet Kirghizia. In several hours the landscape changes dramatically. Around is the richest, most vigorous vegetation, tall cereal grasses growing up half the height of a horse; tremendous areas are covered with wonderful meadows. Rainfall is significantly more abundant here than on the southern slopes of Tien Shan. This is an Eldorado for the nomads. A caravan covers the distance from the Bedel mountain pass to Karakol in several days. We passed by excellent pastures with rich green vegetation. Cattle can feed on these pastures in hundreds of thousands. It is difficult to imagine a more striking contrast than that between Soviet Kirghizia and Chinese Kirghizia beyond the mountain ridge. While in the latter practically all the scanty resources of the desert have already been used, and life is concentrated in a few oases, Soviet Kirghizia, on the contrary, strikes one with the abundance of excellent vast pastures, ample opportunities for cattle husbandry, land cultivation, and an area for developing intensive crops (Vavilov 1965, p. 594-595).

Here rich diversity of invaluable wild species of alfalfa, forage grasses and other plants have been discovered and collected. Such species were considered worth introducing into cultivation. Kirghizia was rich in dyestuff plants, volatile oil plants, opium poppy and new crops. In the regions of Kazakhstan adjacent to Kirghizia a most valuable rubber plant kok-saghyz (*Taraxacum kok-saghus* Rodin) was collected. After Kirghizia the expedition moved to Kashgar Oasis.

The first days alone of exploring allowed Vavilov to define the specific nature of vegetation in Western China. Analysis of cultivated plants represented in this oasis led him to a definite conclusion concerning their undoubted linkage with the vegetation of Central Asia and Fergana. There he found the same Central Asian wheat and barley. However, everything seemed to have undergone, as Vavilov described it, a diminishing process which resulted in making the local plant diversity scantier, so that only a small number of varieties and forms were represented there. Once, the travellers encountered a really fascinating view: in front of them stretched a field which was all covered with white flax with tiny narrow-petalled flowers and white seeds. From the blue-flowered and brown-seeded plant a very peculiar albino had evolved. Yellow and white carrots were also found there (Vavilov 1987b).

The same phenomenon was identified by Vavilov and Popov in wild plants, which showed a very poor representation of species and genera as well as a limited variability in flower colours. The common camel's thorn (*Alhagi sameloru* Fisch) with a red flower had there turned into a form with yellow or even pale yellow flowers.

Their explorations revealed an obviously notable role of isolation and inbreeding. The Kashgar area was separated from Central Asia and India by massive mountain chains, which had been almost impregnable barriers to the distribution of most plants. In the west, the role of the same barrier had been played by the Pamir and Altai Mountains, in the south by the formidable chain of Kunlun Shan and all Tibet, and in the north by the endless ridges of Tien Shan. Thus, this region had been in almost ideal geographical isolation.

Fig. 27. Collecting mission in Xinjiang (1929). N.I. Vavilov is standing fourth in the row with a helmet on [VIR Archives].

When the trip was over, Vavilov was able to state with certainty that Central Asia had nothing to do with the origin of cultivated plants. The latter were introduced there from southwestern Asia or from China. There the distinctive influence of the Iranian and Chinese cultures was obvious.

In October 1929 Vavilov left for Japan. Thanks to his wide recognition in the scientific community, he was welcomed by his Japanese colleagues cordially enough and, with their help, managed to carry out all of explorations he intended to make. Visits to the wonderful Tokyo Botanical Gardens directed by Professor Nogai, the excellent Experimental Station supervised by Dr Kato and Mr Terao, a plant breeder, as well as the meetings with geneticists such as Drs Ikeno, Mayi and others, helped Vavilov to become well acquainted with the agricultural and botanical spheres of the country's life. He was impressed by the endless diversity of plant forms in Japan. Vavilov studied the vast range of plant species and botanical genera which he had never met before with great interest. Among them there were various species of edible bamboo, Chinese yam, enormous and diverse radish, turnip, other root crops, mustard, edible burdock, water chestnut, lotus, arrowhead, Jesuit's nut, edible lily bulbs, extremely diverse and unusual cabbage forms represented by numerous species, unique vegetables called "udo", rhubarb, perennial Chinese onion "jujai", stem lettuce "uisun", original small eggplants, large cucumbers, edible luffa, edible chrysanthemum "shiso", tuberous asparagus and many other crops. Vavilov was also surprised by the Japanese fruit trees represented by such unusual forms as Chinese pear, Japanese and Chinese plum, Chinese cherry, Chinese quince and other fruits (Vavilov 1987b).

Upon reaching Hokkaido, Vavilov was met by Professor M. Akemine, the author of the world's first inventory of cultivated plants now in existence. Together they travelled a lot in the vicinity of Sapporo and to Hokkaido villages. When in Kyoto, Vavilov was the guest of Professor H. Kihara, the famous cytogeneticist. There he studied the large collection of rice assembled by Dr Kato from all over the world. When they went frorn Kagoshima to Sahurajima Island, it was the harvesting season for a radish, called by Vavilov "the masterpiece of the world's plant breeding".

Summarizing his impressions of the crop plants of Japan, Vavilov wrote:

> Collecting and studying of cultivated plants presented real evidence for the absolutely unique nature of these plants which had developed their independent evolution without any influence from the ancient agriculture in south-western Asia. Hundreds of plants are endemic to either China or Japan and many of them still have no wild relatives in China or Japan.
>
> Rice is the main crop planted in the fields and is followed by wheat and barley. Large areas are under citruses, pear-trees and quince, making an unusual rural landscape. Tangerines and oranges take the same place in Japan as apples do in Europe. .....Similiar to the Chinese centre of agriculture, Japan may be characterized by a diverse range of plants, including representatives of the temperate and subtropical climate zones and partially, in the south, from the tropical area. .....The rich diversity of endemic cultivated crop species distinguishes Japan and China among the world's ancient foci of agriculture.
>
> As a rule, these species are represented by an enormous number of different varieties. The diversity of soybean, phaseolus beans, "ajuka", persimmon and citruses is shown literally thousands of easily discernable forms. If one considers the great number of wild plants used in China in addition to cultivated ones, it would be possible to understand to a certain extent why hundreds of millions of people can subsist there (Vavilov 1987b, p. 74-75).

Visiting the Island of Taiwan, which despite its annexation to Japan remained essentially a part of China, was of particular interest for Vavilov. Owing to its isolation from the continent, Chinese agricultural custom was preserved there almost untouched. In Taiwan University, Vavilov was quite courteously received by Professor T. Tanaka, a prominent citrus crop expert. On the same day they worked out a collecting route across the whole island, through the inner territories to the very southern edge. First of all, Vavilov's attention was attracted by the thickets of camphor trees (*Cinnamonum camphora* L.).

At the Tropical Experimental Station in Kagi, Vavilov was acquainted with the plantings of rubber-bearing *Castilloa elastica* Cerv., mangrove and mangosteen, and with excellent collections of gigantic tropical citrus trees and other plant species. Visiting markets and orchards in Taiwan, Vavilov saw an extremely rich diversity of horticultural and medicinal plants belonging to numerous unique species as yet absolutely unstudied. With the help of Professor Tanaka, Vavilov brought from Taiwan many collections of crop seeds, including those of industrial and medicinal plants. Vavilov departed from Taiwan by sea to the Korean peninsula, where upon his arrival in Seoul he developed a plan of explorations there. He intended to make a trip across the peninsula with the aim of studying cultivated plant diversity, gathering as many seeds, scions and other collectable plant materials as possible and achieving an understanding of the peculiarities of Korean agriculture. Cultivated plants in Korea were practically the same as in Japan, mostly rice and soyabean.

Vavilov noticed that the deeper he went toward the central parts of Korea, the more primeval the landscape became. There he was able to observe all the gradual shifts from cultivated forms to the wild ones in their primitive state; it helped him to perceive the genesis of many Chinese cultivated plants. In Korea he saw wild soyabean plants with small seeds and dehiscing pods. Vavilov regarded this form as a wild relative of cultivated soyabean.

Vavilov wrote about his experience in exploring the East Asian region:

> Studying Chinese crop diversity in its marginal areas in Xinjiang, in its locality in Taiwan, Korea and Japan guided us to a certain conclusion that this outstanding culture was wholly unique, that the cultivated plants there were absolutely atypical, that the cultivation techniques were unmatched, and that the ancient agricultural centre in East Asia was completely independent with its uncomon plant species and genera at the basis of their agricultural practice. ...As yet hardly studied, the rich vegetation of China is known only from a few fragments in the reports of European and American travellers and undoubtedly preserves plenty of treasures.
>
> Chinese culture under the peculiar conditions of the monsoon climate had altered the imported wheat and barley forms for thousands of years and created its own unique subspecies. Rice, having originated in India where one can still trace the relationship of this crop with its wild retatives, was brought to China and with a greater degree of cultivation produced unusual varieties..... China was the place of origin of a number of millet-like plants. Sorghum, having come from Africa, was transformed into an uncommon subspecies, kaoliang..... Ahead of us is a vast expanse of work..... detailed examination of plant resources in China and synthesizing the knowledge of these resources. The beaten track of the previous vegetation history should be steered away towards east Asia and southeastern Asia where the most numerous agricultural population had been concentrated for thousands of years (Vavilov 1987b, p. 80-81).

His two months travelling being over, in the last week of December 1929, Vavilov crossed the Korean border and returned to the Soviet Union via Vladivostok. The scientific and practical results of the collecting missions of 1929 were invaluable. He managed to explore an area of 60 000 km², collect approximately 3500 kg of seed and herbarium samples and take several thousand photos illustrating the agriculture and general geography of western China, Japan, Taiwan and Korea.

The Americas

In 1925, the Supreme Council of People's Economy organized a scientific expedition under the leadership of Yu. N. Voronov to South America with the purpose of collecting rubber plants. Vavilov requested that S.M. Bukasov, an expert from the VIR, be included in the team, as the Institute was deeply interested in collecting seeds of the major crops originating in South and Central America, such as potato, Jerusalem artichoke, maize, sunflower, cotton, tobacco, tomatoes, *Phaseolus*, pumpkin, etc. The expedition explored Mexico, Guatemala, Cuba, Panama, Colombia, Venezuela, and parts of Chile and Peru. It continued in 1926, and that year parcels started arriving back. All in all, around 5000 seed samples of various crops were collected. The collected materials of maize, cotton, *Phaseolus*, pepper, pumpkin and other crops made it possible to construct a new taxonomy for them and get closer to the mystery of their origin. The availability of rich initial materials allowed for the implementation of new breeding programmes. Of particular value were the results obtained through a wide ranging study of numerous potato samples organized by Bukasov at the Institute's experimental stations.

The results of the research were presented by Bukasov in 1928 in his monograph *The cultivated plants of Mexico, Guatemala and Colombia* (Bukasov 1928). As Chile and Peru, the most important countries, had not been explored completely, the Institute of Plant Industry sent a second expedition in 1928 headed by Juzepchuk to continue collecting maize, potato, cotton and other South American crops of interest to introduce into cultivation in the Soviet Union. The team of researchers explored Peru, Bolivia and Chile (Vavilov 1980; Soratniki 1984). The detailed results of these investigations were used by Vavilov in his explorations of South America in 1930 and 1932.

In autumn 1930, after participating in the International Conference of Agricultural Economists in Ithaca and the Pan-American Congress of Agriculturists of North and South Americas in Washington DC, Vavilov started implementing the plan of collecting missions in southern states of the USA, Mexico, Guatemala and Honduras. Vavilov's major task was to identify the primary centres of species and variety formation for the major cultivated plants of North America. With that purpose, in addition to visiting agricultural experimental stations in Wisconsin, Illinois, Missouri and Oklahoma, he explored the territories of Florida, Alabama, Louisiana, Arizona, Texas, California, Mexico, Guatemala and some areas in the tropics of Honduras. The results of this collecting trip with an account of Bukasov's and Juzepchuk's explorations were published in 1931, identifying "Mexico and Central America as a basic centre of origin of cultivated plants in the New World".

Proceeding from the materials collected by himself and his predecessors, Vavilov came to the conclusion that the cultivated plants of the New World had comparatively few genera in common with the Old World: for instance, cotton, plum, grapes, apple and hawthorn. The New World before Columbus had been unfamiliar with all genuine cereals, the Old World's leguminous crops – such as pea, chickling vetch, lentil, chickpea – and fodder grasses, alfalfa and clover. Most of the Asian and Mediterranean fruit trees, flax, hemp and most of the Old World's agricultural crops had also been unknown in America before European colonization.

Farming in Pre-Colombian America, as well as crop cultivation, originated in absolute autonomy from the Old World, which is shown by the unique endemic plant diversity of North and South America. Vavilov wrote:

> The basic fact that we can now establish as a result of studies of the entire extensive continent of North America is that there is a striking geographical localization of speciation and variety forming processes with respect to the overwhelming majority of cultivated plants within the continental area of the New World. .....As far as their primary varietal potential, but also that of the wild relatives, are concerned, the great diversity of endemic species of cultivated plants that initially originated in North America turned out to be associated with the exceptionally limited territory of Central America, consisting of southern Mexico, Guatemala, Honduras, El Salvador, Nicaragua, Costa Rica and Panama. However, it is necessary to subtract a considerable portion of the large Yucatán Peninsula from this comparatively small territory (Vavilov 1960, p. 139).

According to Vavilov's data, a considerable part of global plant resources derived from this region of Central America. This was the place of origin of maize, upland cotton, several species of gourd, *Phaseolus* beans, chayote, cocoa, henequen, papaya, several ornamental plants, such as dahlias, cosmos, zinnias, tagetes and morning glory. The edible

Fig. 28. N.I. Vavilov in the market place in Venezuela, 1930 [VIR Archives].

Fig. 29. N.I. Vavilov in Mexico, 1930 [VIR Archives].

potato originated in South America. It came from Chile[4] though in Mexico and Central America there are over 30 wild potato species. Only sunflower and Jerusalem artichoke are attributed to the present-day territories of the USA and Canada.

Further, Vavilov wrote:

> A great wealth of endemic species of cultivated plants can be found in south Mexico. The greatest racial diversity of maize is still concentrated just within this area and only there one can still see a large amount and variety of teosinte (Vavilov 1960, p. 145).

In the same area he found a great diversity of maguey agave, yam beans, tomato (*Lycopersicon cerasiforrne* Dunal.), scarlet runner beans, large-grained maize varieties, *Cucurbita moschata* Duch, *Parthenium argentatum* A. Gray, *P. incanum* H.B.K., cacao and vanilla.

In Vavilov's opinion, such regions as Honduras and Guatemala had not by that time undergone serious plant exploration. During his trip, he could not pay enough attention to this aspect. Vavilov believed that most of the New World's cultivated plants imported to the Old World had originated in Central America and Southern Mexico, while many endemic species from Colombia, Peru, Chile and Brazil, such as *Ullucus tuberosus* Loz, *Oxalis tuberosa* Mol., *Chenopodium quinoa* Willd. and others, had not yet been widely used in the Old World.

---

[4] Now considered to have evolved in the Central Andes of Peru and Bolivia.

In South America, Vavilov marked out the Peruviàn Centre of diversity. He wrote:

With respect to the number of endemics, the Peruvian centre is comparable to the south Mexican and Central American centres. However, in contrast to Cook (1916, 1925), we consider the latter to be more important as far as cultivated plants are concerned with respect both to amount and composition (Vavilov 1960 p. 156).

In conclusion, Vavilov stated:

The investigator can only begin the study of plant resources in Central America and South America while, so far, making use of the crops of the primitive farmers; they mainly used plants within the limits of the tropics and the mountain areas where the first civilizations of ancient America settled.....The enormous stores of plants that are concentrated in the tropical regions of Central America and South America are still not fully known to man. Some 10,000 species are characteristic of the floras of Brazil, Peru, Venezuela and Colombia. Primitive man avoided these regions; the powerful forces of tropical nature stopped him. He fought and to this day still fights the diseases associated with the terrible elements of the Tropics. However, during the twentieth century and in the future, the intention is to master the Tropics and reveal the colossal plant resources that can be utilized for the various needs of mankind (Vavilov 1960, p. 158).

In 1932 after the Sixth International Genetical Congress in the USA, Vavilov started implementing a programme of germplasm collecting over the American continent. Vavilov wrote about his impressions of this trip in his letters:

After Canada and the USA, Vavilov visited Puerto Rico, Cuba, the Yucatán Peninsula in Mexico, the western coast of Central America, as well as Ecuador, Peru, Bolivia, Chile, Argentina, Uruguay, Brazil and the Island of Trinidad. In his letter to N.V. Kovalev mailed in Peru on 7 November 1932, Vavilov wrote:

Studying flowering potato fields in Peru has convinced me that all the so called local varieties can still be split into hundreds of forms..... In other words, there are millions of botanical varieties and forms. Our ignorance concerning the Andean potato diversity is striking..... There is a damned multitude of wild species, but the cultivated potato is such as I have never seen before, being still unacquainted with "the furnaces of creation". And everything here is connected with wild materials..... I have been in Yucatán and now have a more or less full concept of the whole of South America..... I am collecting everything...It is quite clear with maize: Central America. I think that with cotton we have the maximum interest in Central America as well. I have secured Peruvian endemics.

I have sent 6 parcels of 5 kg each. I cannot help sending them..... Tomorrow I will be at the Bolivian border..... In the Cordilleras I am searching for the essence of crops and varieties; in a month I will study the future of the world's agriculture in Argentina and Brazil. I am in a hurry (Vavilov 1987a, p.185).

Explorations in North, Central and South Americas from August 1932 to February 1933 resulted in the delivery to the Soviet Union of numerous parcels of new plant materials. In Argentina a complete set of cereal breeding cultivars was acquired, together with the best flax, maize and wheat varieties bred in recent years, quinine seeds that made it possible afterwards to establish plantations on the Black Sea coast of the USSR, large numbers of new

Fig. 30. N.I. Vavilov in Arizona, USA, 1932 [VIR Archives].

Fig. 31. N.I. Vavilov and Dr Henry, Botanical Gardens, Buenos Aires, 1932 [VIR Archives].

Fig. 32. Third from the left: N.I. Vavilov at the Experimental Station in Pergamino, Argentina, 1932 [VIR Archives].

varieties and species of cultivated and wild cotton that helped to attain excellent progress in cotton breeding in the Middle Asian Republics and Transcaucasia, a wide assortment of fodder grasses including new sweet clover varieties suitable for the production of green fertilizers in the Subtropics of the USSR, new potato materials, and a series of disease-resistant cereal varieties. For the first time a number of new cultivated plant forms endemic to Peru, Bolivia, Ecuador, Salvador, Brazil and Trinidad were collected and delivered. From all the countries explored Vavilov gathered over 2000 publications of local agricultural and botanical literature.

Other travels

Not all of Vavilov's plans were fulfilled; for example, he had planned to visit Egypt, Sudan, India and Iran. He wrote about that in a number of letters to his co-researchers and followers between 1932 and 1935:

> I would like to go [in 1932] for a month to Egypt and Sudan in order to get acquainted with agricultural practice and visit the leading scientific institutions, especially those dealing with cotton production. Cultivated cotton in Egypt now represents a great deal of interest for us, since we have started working with cultivated Egyptian cotton (Vavilov 1987a, p.171).

> All my concern is now [1933] with India and South-East Asia; I am summarizing the worldwide plant philosophy and distribution for we have logically approached this very point, while all other issues are more or less clear..... As for me, last year I was deep in and this year I am finished with Central and South America; we know quite a bit about South West Asia. On the whole it is necessary to begin serious studies of India, Indo-China and China (Vavilov 1987a, p. 206).

> This autumn [1934] I shall be unable to go to Persia myself, as I have been commissioned to restructure the Academy of Agricultural Sciences. But the business should not be put off. I would myself request you to keep in mind my trip at a modest scale in the next year (Vavilov 1987a, p. 240). [And, in 1935] Our interests in India are growing from year to year. These days I shall send you a book with a map where our work is generalized, and it will help you to understand why our eyes are fixed on India (Vavilov 1987a, p. 268).

In 1937, Vavilov intended to undertake an expedition to southeast Asia and make the Philippines the first place to visit. Seed samples were collected in foreign countries, not only by the collecting missions of Vavilov and his collaborators. Other opportunities were also used, including business trips of staff and their visits to various plant breeding and seed-production enterprises (Vavilov 1980, 1987a; Soratniki 1994). For example, in 1927-1929, F.D. Likhonos, a fruit breeder, was visiting research institutions of Germany, Austria, Yugoslavia, Italy, France, USA and Canada which resulted in delivering over 1000 samples of new fruit crop varieties, scion wood of the best commercial apple cultivars as well as cereal, small grain, vegetable, industrial and forage crop samples. In 1930, about 200 samples of new commercialized varieties of all agricutural crops were brought by N.N. Kuleshov from his trip to the United States where he was studying the conditions and characteristic features of maize and sorghum cultivation. In 1936, after the visit to Japan of G.N. Shlykov, Head of the Plant Introduction Department, the collection was enriched with approximately 1500 samples of cereal, leguminous, forage, ornamental, fruit and citrus crops. Replenishment of the collections was accomplished by requesting germplasm by mail, and by researchers collecting plant samples from other botanical and crop production institutes (Vavilov 1980).

Fig. 33. Expedition in Turkmenia, 1936. N.I. Vavilov is sitting in the centre [VIR Archives].

In 1924-1925, a worker of the Main Botanical Garden of the USSR Academy of Sciences, Ms E.G. Chernyakovskaya, brought approximately 1000 samples of wheat, barley, small grain crops maize, vegetables, melons and legumes from Khorasan (northeastern Persia) and from Seistan in central Persia, a region previously unavailable for studying. In 1925, a collection of leguminous samples was received from Abyssinia, Burma and India. In 1926, the experimental institutions in Bulgaria submitted most valuable materials of the local wheat varieties collected in the almost inaccessible mountainous areas of the country. In 1929, extremely interesting cotton and flax samples collected near Kashgar were received from M.F. Dumpis, Consul General of the USSR in Afghanistan, whom Vavilov met in 1924 during his visit to the country. In 1930, the Institute received early-ripening plant samples from the mountainous parts of Arabia and Yemen. In 1931, Professor Tanaka, who had accompanied Vavilov around Taiwan, sent Vavilov new local citrus samples. In 1935, following Vavilov's request, the Central Experimental Station in Nanking, China, sent a specially prepared collection of wheat samples numbering 400 varieties from various regions of China.

In the 1930s, Vavilov had been devoting a lot of time to the questions of the development of the evaluation of rich plant diversity and perspectives for its rational utilization in the Central Asian republics. Vavilov visited Tadjikistan for the first time in 1916, and then again in 1924. In the 1930s, he had repeatedly visited the republic. In the work *Cultivated flora of Tadjikistan, its past and future* (1934) Vavilov wrote that:

> .....the indigenous flora of Tadjikistan is extremely rich in species, and despite the comparatively limited territory, represents one of the most interesting regions within the USSR, comparable only with some Transcaucasian regions (Vavilov 1965).

Vavilov noted that no less than 4000 species of flowering plants occur in Tadjikistan. The region offers ample opportunities to search for new plants (i.e. fruits, rubber, volatile oil, medicinal, industrial plants, etc.). In the above-mentioned work, Vavilov provided a thorough analysis of plant industry in Tadjikistan and indicated trends in its development.

Vavilov went to Turkmenistan, as well as to Tadjikistan, for the first time in 1916 in order to explore cultivated plants along the Tedzhen and Atrek rivers and in the Murgab river valley. Later, in 1925, he explored the lower Amu Darya river. In the 1930s, he visited the Turkmenian Experimental Station in Kara-Kala, and the Repetek Sandy Desert Station several times. In 1935, in his report entitled *Agricultural Turkmenia,* Vavilov expressed his ideas regarding the present and future of plant industry in Turkmenia. In that report, he wrote about prospects for the cultivation of cotton, forage crops, cereals, melons, vegetables, fruit crops, viticulture development and the introduction of new crops, as well as about the status of wild fruits in Turkmenistan.

Vavilov visited the Caucasus several times, beginning from his trip when a student until the end of his plant-exploration activities (1928, 1933-1936 and 1939). In 1939, Vavilov took part in the Caucasian Complex Expedition of the Academy of Sciences of the USSR as leader of the Agricultural Group. For three months, from July until September, teams of the Group had been carrying out a planned exploration of all of the mountainous regions of the Krasnodar and Stavropol Territories, Cherkess Autonomous Region, Kabardino-Balkarian and North Ossetian Autonomous Soviet Socialist mountainous regions of the Caucasus. Some results of the Caucasian expedition were sketched by Vavilov in early 1940, but they were published only 17 years later in the paper *Mountainous agriculture of the North Caucasus and prospects of its development* (Vavilov 1957b). During that expedition Vavilov discovered a new rye subspecies, *Secale cereale* subsp. *dighoricum* Vav., a field weed with fragile ears. This rye, discovered in Dighoria (North Ossetia) in the Urukh river canyon at altitudes from 1350 to 1850 m asl, was a strong weed in fields of spring barley and wheat, and to a lesser extent, winter wheat. The feature of outstanding ear fragility was reminiscent of the wild rye species *Secale montanum* Gus. and *S. fragile* MB.

Before 1940, VIR researchers carried out systematic collecting of plant resources in Karelia, the Kola Peninsula, Byelorussia, Ukraine, the central black-soil regions, the Lower Volga regions, Siberia and Kazakhstan. In the first place, thorough explorations and sample collecting were carried out in the Central Asian region, Transcaucasus and the Far East as, according to Vavilov, these were the regions of the country most rich in specific and intraspecific diversity. These regions donated to the VIR collections nonligulate forms of wheat and rye occurring nowhere else in the world, wild forms of almond, fig, apricot, wild volatile oil and rubber plants, numerous forms of forage grasses, wild pear, steppe cherry, etc. In addition, landraces of various crops from these regions were added to the collections at VIR.

From 1923 to 1940, VIR carried out 180 collecting missions, of which 140 explored the territory of the Soviet Union. Consistent introduction activities by Vavilov and his researchers were concentrated at first on grain, industrial, vegetable, fruit and other groups of cultivated plants and their wild relatives. Later, special attention was given to such new crops as tung, jute, rubber plants, and some medicinal plants.

The number of collected accessions illustrates the rate of introduction. For instance, the collections numbered over 36 000 accessions of wheat, over 10 000 of maize, over 23 000 of legumes, around 18 000 of vegetables, over 12 000 of fruit and small fruit crops and over 23 000 of forages. The total number of accessions reached 250 000 in the time of Vavilov. All this rich diversity was thoroughly studied at experiment stations in different geographical zones of the country (Vavilov 1965).

Fig. 34. Expedition in Central Asia, 1936. N.I. Vavilov is in the front [VIR Archives].

Fig. 35. At the Near Aral Experimental Station of VIR, 1936. N.I. Vavilov is second from the left [VIR Archives].

5 Vavilov's theoretical contributions

The central issue which Vavilov worked on throughout his life was the concept of the global genetic diversity of cultivated plants. This concept incorporates a series of his fundamental theoretical generalizations that outlined new directions in the theory of plant introduction and applied botany, won worldwide fame for the national science, and played a significant role in the development of genetics and plant breeding in the Soviet Union and abroad. The theoretical bases of this concept included the law of homologous series in hereditary variation, fundamental studies on the problem of the species as a system, botanical and geographical principles of plant breeding, and the key theory of the centres of origin of cultivated plants.

The Law of Homologous Series in hereditary variation

The Law of Homologous Series has great importance from the standpoint of appreciating the diversity of plant forms and their efficient utilization. It has set a distinct framework for the taxonomy of cultivated plants and provided botanists, plant researchers, geneticists and breeders with a clearer idea of the place of each taxon in the immense treasury of the plant world. This law helped to discover new plant species and forms and aided the experimental production of new plant types, which proved to be interesting for plant production and breeding.

The law became known in three versions: as a brief report in 1920 at the Third All-Russian Plant Breeding Congress in Saratov, as a comprehensive publication in the *Journal of Genetics* in 1922, and as a chapter in *Theoretical Bases of Plant Breeding* in 1935. The latter was a substantially supplemented version of the concept taking critical comments into account. Although some reviewers stated that it would have no practical value, the contrary very soon proved to be the case. This theory helped to give a definite basis for collecting agricultural crops, enabled a scientifically based strategy for such collecting to be worked out, and laid the scientific foundation for the famous global plant collections of VIR.

In 1920, in his letter to L.S. Berg written in Saratov, Vavilov said:

> All the attention of our laboratory is being paid to the search for series in isolated genera which cannot be intercrossed. Vetch, lentil and pea cannot be crossed, but their series of variation, we may say, are almost the same. These days we have received new vetch samples from Kharkov, so all the gaps in the series are now filled. For 3 years we have been trying to do the same with other forms of cultivated plants, but this is difficult, as we have got too many characters and hence it becomes very cumbersome. We have reached the formulae but this is still unsatisfying. This summer we replenished the series of Cucurbitaceae; most complete are the series in Cruciferae, *Eruca* and *Brassica* which have shown identical series for a lot of characters, even down to anatomical structure. We have not got enough examples. It is necessary to attract everything existing in the world. I shall do all I can to send somebody abroad in the autumn of 1921 to collect plant materials. We need to delegate a collecting mission to Africa where cultivated plants have practically never been studied. I am sure that new works would unveil the mystery of the unknown, although it is almost a Utopia in terms of time (Vavilov 1980, p. 41).

In his paper in the *Journal of Genetics,* Vavilov wrote:

The detailed study of variation among many different groups, and the great number of new facts permits us to take this subject anew and to bring all known facts into the form of a general law to which all organisms are submitted... (Vavilov 1922a, p. 53).

Thus, after investigations of local Russian and Asiatic wheats at our experimental station, the existence was shown of about 3000 Jordanons (botanical varieties) of *Triticum vulgare* Vill., perfectly recognizable morphologically and physiologically. For barley we know at least 600 to 700 Jordanons, for oats more than 600. In Rye, *Secale cereale*, many hundreds of forms, differing in hereditary morphological and physiological characters, were collected by Mrs V.P. Antropova, from different parts of Persia, Bokhara, Asiatic and European Russia... (Vavilov 1922a, p. 49).

There is no essential difference in this respect between wild and cultivated plants... (Vavilov 1922a, p. 50). Species and genera that are genetically close are characterized by similar series of heritable variations with such regularity that knowing the series of forms within the limits of one species, we can predict the occurrence of parallel forms in other species and genera. The more closely related the species and Linneons (species) in the general system, the more resemblance will there be in the series of variations.Whole families of plants in general are characterized by cycles of variability occurring through all genera and species making up the family... (Vavilov 1951, p. 75).

The multitudinous chaos of innumerable forms obliges investigators to look for some way of simplification. The process of differentiation will go on inevitably, adding to the records of existing forms, and giving a true conception of Linneons. But parallel to differentiation it is natural to search for ways of integration of our knowledge of Jordanons and Linneons themselves... (Vavilov 1922a, p. 52).

Thus the Linnean species, in our conception, appears to be a distinct, complex, mobile, morpho-physiological system related in its origin to a definite environment and area, and in its intraspecific hereditary variability, subject to the Law of Homologous Series. The Law of Homologous Series shows investigators and breeders the directions for their search. It aids in the discovery of systematic links and extends the horizon of the worker, disclosing the great amplitude of species variations... (Vavilov 1951, p. 91).

The problem of the origin of species cannot be separated from the problem of variation. A great many forms are undoubtedly only different combinations of the same genes, some primary types. The study of variation will give us the possibility of establishing these primary types, the fundamental series of variation of organisms (Vavilov 1922a, p. 83).

Centres of origin

The first research paper by Vavilov concerning the problem of origin of cultivated plants was published in 1917. It was entitled *On the origin of cultivated rye* and based on the results of the first collecting mission to Persia and the Pamir mountains. The second one, *On the eastern centres of origin of cultivated plants,* appeared in 1924 after explorations in Mongolia and Transcaucasia, and the study of materials from Asia and Africa. All the new materials and

Fig. 36. Homologous series in hereditary variation. Example of the variation in the ear of wheat species [VIR Archives].

information regarding the distribution of plant resources around the globe brought by collecting missions headed by either Vavilov himself, or by his colleagues found reflection in Vavilov's works (Vavilov 1917, 1924). In 1923, while working on the paper *On the eastern centres of origin of cultivated plants* (Vavilov 1924), Vavilov wrote in a letter:

> It became possible to clarify the question of the centres of origin of cultivated plants. For instance, there turned out to be two centres for barley – in eastern Asia and in eastern Africa; for flax – in North Africa and southeastern Asia. The centres for crops and the centres of origin of cultivated plants turned out to be located not along the valleys of the great rivers, as I thought before, but on the contrary in the mountainous regions (Vavilov 1980, p. 136).

In 1926, Vavilov presented a fundamental concept *The centres of origin of cultivated plants* in the *Works on Applied Botany and Plant Breeding*, and paid tribute to A. De Candolle by dedicating this work to the great investigator of cultivated plants. In this publication Vavilov summed up results of theoretical investigations and, proceeding from the Law of Homologous Series in Variation, stressed that formation of various genera and families is characterized by parallelism of a cyclic nature which makes it possible to predict the existence of certain forms and therefore simplify to some extent the solution of the question of their origin (Vavilov 1926b, 1992).

When summing up for the first time the results of his theoretical investigations, Vavilov singled out five major foci of the main field, garden, and orchard crops, saying that:

> ...in addition to the centres, we will, in the future, certainly be able to outline a number of secondary centres and to specify more the actual geographic centres... The areas of origin and type-formation of the most important cultivated plants which, at the same time, are the foci of a wealth of types, belong mainly to the mountain areas of Asia (Himalayas and its system), the mountain systems of northeastern Africa and the mountain areas of southern Europe (the Pyrenees, the Apennines and the Balkans), the Cordilleras and the Southern spurs of the Rocky Mountains. In the Old World the original areas of cultivated plants belong mostly within a belt between 20′ and 40′ N latitude (Vavilov 1965, p.100).

Such a diversity of conditions ranged from deserts to oases, from poor stony soils to the humus-rich alpine and subalpine meadows, the abundance of irrigation water from the melting snow and ice, the isolated location of those areas and their safe position regarding invasions – all of which Vavilov thought facilitated the accumulation and development of an exceptional plant diversity.

Vavilov wrote:

> The region of maximum variation, usually including a number of endemic forms and characteristics, can usually also be considered as the centre of type-formation.The elucidation of the centres of type-formation and the origin of cultivated plants allow us to approach objectively the establishment of basic foci of agricultural civilizations... Plants and their varieties are not easily dispersed from one area to another; in spite of the many thousands of years of wandering about by peoples and tribes; it was, as we have seen, not difficult to establish the basic foci of type-formation of the majority of cultivated plants. The presence in northern Africa and south-western Asia of large groups of endemic plants, both species and varieties of cultivated plants, on the basis of which independent agricultural civilizations arose, can solve the problem of the autonomy of these civilizations, also from the historical point of view (Vavilov 1965, p. 102).

Vavilov concluded as follows:

> Apart from the spontaneous and utilitarian importance of understanding the sources of a multitude of plant types, the ultimate objective of the investigations discussed is to try to approach, in earnest, the general biological problem of speciation. Evolution occurs both in space and time. It seems to us that only by earnestly approaching the geographical centres of type-formation and by establishing all the links connecting the species will it be possible to find the way to the establishment of a system of Linnaean species and to understand the latter as a system of forms... Hence, the very problem of speciation becomes a problem concerning the evolution not only of different races, which, according to Darwin's hypothesis, were isolated into basic species, but also of the origin of the complicated system within which the Linnaean species developed.
>
> As a result of all that has been said above, the solution of the problem of speciation lies in a synthesis of a far-reaching investigation of various groups of plants, while using the methods of differentiating taxonomical phytogeography for the purpose of establishing centres of type-formation, and the methods of genetics and cytology. The way to grasp the integrity of the species can only be found in a synthesis of taxonomy, differential geography, genetics and cytology (Vavilov 1965, p. 103).

Vavilov regarded this work as the initial stage of investigations, the further development of which could help to specify and more clearly define the borders of the centres, or foci, of origin of cultivated plants. For more than two decades Vavilov worked extensively on this problem, as he regarded it as the most important. Thus, in 1927 he published *Geographical regularities in relation to the distribution of the genes of cultivated plants, The problem concerning the origin of cultivated plants as presently understood* (Vavilov 1927b, 1992) and *Geographical localization of wheat genes on the globe* in 1929; *The role of Central Asia in the origin of cultivated plants, Mexico and Central America as a basic centre of origin of cultivated plants in the New World* (Vavilov 1992) and *Wild relatives of fruit trees of the Asiatic part of the USSR and the Caucasus, and the problem of fruit tree origin,* as well as *The problem concerning the origin of agriculture in the light of recent research* in 1931; *The phytogeographical basis for plant breeding* (Vavlov 1951, 1992) in 1935; *The important agricultural crops of pre-Columbian America and their mutual relationship* in 1939, and finally *The theory of the origin of cultivated plants after Darwin* (Vavilov 1992) in 1940.

In the paper entitled *The phytogeographical basis for plant breeding* published in a collective fundamental work *Theoretical Bases of Plant Breeding,* Vavilov wrote:

> Our initial aspirations were directed mainly toward the study of difficult objectives such as wheat, rye, barley, maize and cotton, which are at present widely grown all over the world and have already long since been dispersed from the primary centres, where they were initially taken into cultivation...
>
> As far as the introduction of new objectives into the study is concerned, it became increasingly evident that there is a coincidence between the areas of primary type-formation of many species and even genera. In a number of cases, literally dozens of species could be referred to one particular area. The geographical studies led to the establishment of entirely independent floras of cultivated plants, that are specific for different areas (Vavilov 1960, p. 28).

> When summing up the work done by the Soviet teams of plant breeders and the expeditions sent out over many years to Asia, Africa, Southern Europe and Central and South America and covering up to 60 countries, and when reviewing the results of the detailed comparative studies of the colossal amount of new material of varietal and specific diversity collected, we have been able to establish eight independent centres in the world, where the most important cultivated plants originated.
>
> The work in this direction is not yet completed: we still know southeastern Asia very inexactly and a number of expeditions are still necessary for China, Indo-China and India in order to be able to define more exactly the centres of original type-formation of their cultivated plants and to be able to understand the new material. However, we can speak with considerably greater accuracy than dreamed of ten years ago about the eight ancient and basic centres of agriculture in the world, i.e. more accurately about the eight independent areas where the plants were initially taken into cultivation. In our previous papers we delimited the centres of agricultural origins through a study of some basic indicator-crops. There are not enough data for an exhaustive approach. In the present list we have tried as far as possible to provide a complete review of the crops typical of the different areas. We have made serious changes in and additions to our previous ideas, which were initially expressed in the book on the "Centres of origin of cultivated plants" in 1926. Most of the expeditions and most of the work, while studying the varietal resources of the world, were conducted during the period from 1923 to 1933 (Vavilov 1960, p. 29).

In this paper Vavilov had already described eight major ancient focal centres of global agriculture:

I.	Chinese	VI.	Abyssinian
II.	Indian	VII.	South-Mexican and Central American (including the Antilles)
IIa.	Indo-Malaysian		
III.	Central Asian	VIII.	South American (Peruvian-Ecuadorean-Bolivian)
IV.	Asia Minor	VIIIa.	Chilean
V.	Mediterranean	VIIIb.	Brazilian-Paraguayan

For each of the centres or foci Vavilov listed major species of cultivated plants characteristic of the particular geographic region. This list included cereal grasses, grain legumes, bamboos, root crops, tuber crops, bulb and aquatic food plants, vegetable, melon, fruit, forage, sugar, oil-bearing, volatile oil, resiniferous and tanning plants, spices, industrial and medicinal plants, fibres, dye plants, multipurpose plants, and endemic plants.

In 1940, in *The theory of the origin of cultivated plants after Darwin* Vavilov wrote:

> Asia appears to be the area furnishing the highest number of cultivated plants. Its share of the 1,000 species concerned is about 700, that is to say around 70% of all cultivated plants. Approximately 17% have arisen in the New World. Before the arrival of the Europeans, Australia had no cultivated plants and it is actually only during the last century that its eucalyptus and acacia trees have begun to be widely grown within the tropical and subtropical areas of the world. The following seven main geographical centres of origin of cultivated plants can be distinguished within the borders of the continents:
>
> 1. The South Asiatic tropical centre embraces the territory of tropical India, Indo-China, southern China and the islands of southeastern Asia... Within this large geographical centre or area, three foci can be distinguished, which are considerably different with respect to the complex of cultivated plants which are characteristic of them.
>
> (a) The Indian focus (with the richest cultivated flora);
> (b) The Indo-Chinese, including Southern China; and
> (c) The island focus, including the Sunda Islands, Java, Sumatra, Borneo, the Philippines, etc...
>
> 2. The East Asiatic centres including the temperate and subtropical parts of eastern China and the major portions of Taiwan, Korea and Japan... Within this centre can be distinguished a main Chinese focus and a secondary, mostly Japanese one.
>
> 3. The southwestern Asiatic centre covers the interior foothill area of Asia Minor (Anatolia), Iran, Afghanistan, inner Asia and northwestern India. The latter is floristically linked to Iran (as far as cultivated plants are concerned). It also covers the Caucasus, the cultivated flora of which is genetically associated with that of Asia Minor. This centre can be distinguished into the following foci:
>
> (a) The Caucasian focus with the majority of endemic species of wheat, rye and fruits. As explained by comparative cytological and immunological investigations, this is the most important focus of origin in the world as far as wheat and rye are concerned.
> (b) the Asia Minor focus, including the interior portion of Asia Minor (Anatolia), interior Syria and Palestine, Transjordania, Iran, Northern Afghanistan and Inner Asia (including Chinese Turkestan); and
> (c) the Northwestern-Indian, in addition to the Punjab and the adjacent provinces of Northern India also covering Belutchistan, southern Afghanistan and Kashmir...

4. The Mediterranean centre embraces the countries situated along the coasts of the Mediterranean Sea...In addition, it is still possible to trace the close association between the origin of various crops and particular territories such as, e.g., the Iberian peninsula, the Apennines, the Balkan peninsula, Syria and Egypt. Each of these foci is typical of original species of forage plants...

5. Within the borders of the African continent, the tiny Abyssinian centre can be distinguished as an independent geographical area... To this centre belongs also the somewhat peculiar Arabian mountainous or Yemenite focus, which reflects influences from both the Abyssinian and the southwestern-Asiatic centres and is characterized by extremely fast-ripening kinds of cereals, leguminous plants and alfalfa.

6. A Central American geographical centre can be particularly distinguished within the wide-ranging territory of North America. It embraces Southern Mexico as well and can be subdivided into three foci:

(a) the mountainous South-Mexican focus;
(b) the Central American; and
(c) the West Indian focus...

7. The Andean centre within South America covers parts of the Andes. We can distinguish three foci in that area:

(a) The Andean focus proper, including the mountain area of Peru, Bolivia and Ecuador...
(b) The Chilean (Araucarian) focus, situated in Southern Chile and on the neighbouring islands, gave rise to the species of the common potato, *Solanum tuberosum*. In contrast to the Peruvian, Bolivian and Ecuadorean potatoes, which usually form tubers during the short equatorial days, the common potato does so in the conditions of medium-long days at a latitude of 38-40 S in Southern Chile...
(c) The Bogotá focus in Eastern Colombia was established by the Soviet botanists, Drs S.M. Bukasov and S.V. Juzepchuk. The crops are raised at a high altitude (up to 2.800 m elevation).

Actually, the seven large centres distinguished correspond to the sites of ancient agricultural civilizations. The tropical South-Asiatic centre is associated with the high level Indian and Indo-Chinese ones. The most recent excavations reveal the great age of these cultures, which are contemporary with those in Asia Minor. The East-Asiatic centre is associated with the ancient Chinese civilizations and the Southwestern-Asiatic centre with the ancient civilizations of Iran, Asia Minor, Syria and Palestine. The Mediterranean centre was, several centuries before the present, already associated with the Etruscan, Hellenic and Egyptian civilizations, which are estimated to have existed about 6,000 years ago. The comparatively primitive Abyssinian civilization has deep roots, and was contemporary with the Egyptian one or perhaps even pre-dated it. Within the New World, the Central American centre is linked to the great Mayan civilization, already in existence before Columbus, which was enormously successful in science and the fine arts. The Andean centre is associated with the remarkable pre-Inca and Inca civilizations (Vavilov 1965, pp. 164-169).

This paper summing up results of work carried out by a large group of researchers headed by Vavilov appeared to be the last in a series of publications devoted to the problem of the origin of cultivated plants. In it, using an enormous amount of factual material, Vavilov presented in full detail his concept of the centres and foci of origin and type-formation of cultivated plants.

The study of regularities in geographical distribution of the Earth's plant resources and the establishment of tremendous interspecific diversity in most crops made it possible not only to determine the place, but also to make a judgement of the time of origin of the major cultivated crops:

> The history of the origin of human civilizations and agriculture is, of course, much older than the documentation, in the form of relics, inscriptions and bas reliefs can tell us about the past. The more recent acquaintance with cultivated plants and their differentiation into geographical units compels us to attribute the very origin of these to distant epochs where periods of 5-10,000 years mean only a short time (Vavilov 1992, p.12).

The method of differential taxonomy of several hundreds of cultivated plants made it possible to trace their movement, and to identify stages of evolution and introduction into cultivation of domesticated plant species. Studies on the genesis of separate cultivated plants led Vavilov to the new idea of more ancient, primary and less ancient secondary crops which allowed him to characterize with great precision the foci of origin of cultivation, and routes of crop migrations.

The number of centres in Vavilov's works, issued over a comparatively short period varied from three in 1924 to five in 1926-1927, six in 1929-1930, seven in 1931, eight in 1934-1935, and again seven in 1940. Each work was a result of a thorough analysis of new facts. In the works of 1934-1935 it was proposed to divide the southwestern Asiatic centre into Middle Asiatic and Minor Asiatic ones. In 1937 the Middle Asiatic centre was given the name of "Central Asiatic centre", belonging to one of the five most important regions of origin of cultivated plants in Asia. It embraced northwestern India, Afghanistan, Uzbekistan, Tadzhikistan, and part of eastern Turkmenistan. Vavilov explained the further abolition of the division of the southwestern Asiatic focus by the similarity in the specific composition of cultivated flora in that territory.

His ideas regarding the Mediterranean and Abyssinian centres did not undergo significant change after 1926 when they were defined. Botanical explorations in the American continent carried out by Vavilov, Bukasov, Juzepchuk and other researchers from the Institute led to the identification of autonomous foci of origin of agriculture based on species and even genera unusual in the Old World. The comparative youth of civilizations in the New World makes it possible to observe the separation of cultivated species from their wild relatives. Within the New World, Vavilov identified two geographical centres with a striking localization of type-formation in cultivated plants, namely the Central American and South American centres. The latter was named the Andean centre in 1940.

In all his works, starting with *The centres of origin of cultivated plants,* the most fundamental, Vavilov used the terms "centre", "focus", or "area" (of origin) as equivalents for denoting the locations of ancient agricultural civilizations. Of significant importance are such definitions as "geographical centre", "major autonomous focus of origin of agriculture", "geographical areas of concentration of cultivated plants". Proceeding from one of Vavilov's works devoted to the origin of cultivated plants to another one, it can be seen that the terms "centre" and "focus" are attributed more and more to vast territories.

In connection with this, in his last works Vavilov wrote about the the conventional nature of the concept of "the centre" of origin used by Darwin. His work *The theory of origin of cultivated plants after Darwin* was published in 1940, and was dedicated to this problem. In this work he made clear the distinction between "centre" and "focus". According to Vavilov, "centres" are the seven centres of crop origin, which are divided into a number of type-formation foci. It should be emphasized here that Vavilov set forth brand new ideas.

According to his theory of centres of crop origin, the diversity of cultivated plants originated and developed in a few areas situated usually in mountainous areas. They were the primary centres of polymorphism. As for river valleys, which housed great ancient civilizations, they were only the zones of agricultural development, representing secondary centres of type-formation of cultivated plants. However, according to Vavilov, one and the same region could well be the centre of origin of one group of crops and the secondary centre of polymorphism of other crops. Besides, he also distinguished autonomous centres of type-formation of cultivated plants.

After his explorations in the Mediterranean countries and Ethiopia, Vavilov formulated the regularities in geographic distribution of genes, and later found more proof for them. In his report at the Fifth International Genetic Congress held in September 1927 in Berlin and later in his publication *Geographic Regularities in Relation to the Distribution of Genes of Cultivated Plants* Vavilov summarized the results of the study of the materials collected by him. Reflecting over the geography of plant gene distribution, he put forward the idea that dominant genes of any cultivated plant species are concentrated in the centre of its origin, while recessive ones manifest themselves at its periphery. With this, in some isolated territories, even within the limits of centres of origin, it was possible to find populations with high concentrations of separate recessive characters. Vavilov's observations showed that if dominant genes predominated in plants in primary centres of origin, many recessive traits revealed themselves in a natural way on the boundaries of the areas of distribution of plant species. Thus, secondary centres of type formation were also the sources of the crops with valuable recessive genes for plant breeding. The idea of geographic distribution of gene alleles, which matched by its value the law of homologous series in hereditary variation, was of great practical importance for breeding.

Within a short period of time, Vavilov and his researchers made significant corrections to De Candolle's concepts regarding the origin of the major cultivated plants, and determined the origin of several hundreds of others. These data, however, were for Vavilov just raw material on the way from the particular to the general – in determining the world's major centres of origin.

Disease resistance

While solving the problem of the origin of cultivated plants, Vavilov looked at a series of other issues having importance for biology in general, including the problems of species and type formation, and of resistance. In the course of his life, he applied great effort and much attention to the genetics of immunity. His first publication on this issue appeared in 1915. In 1919, the book *Plant immunity to infectious diseases* was published, and in 1935 the monograph *Study of immunity of plants to infectious diseases (with reference to plant breeding requirements)* appeared (Vavilov 1951). Finally, in 1961 his posthumous publication *Regularities of natural immunity to infectious diseases (clues to finding immune forms)* appeared.

In these works he substantiated the concept of the genetic nature of immunological phenomena, revealed the role of pathogen specialization, and the linkage of immunity to ecogeographical groups, as well as addressing many other problems. He demonstrated that immune reactions of a host to a pathogen were determined by the host's genetic nature. Also important is the existence of physiological races of fungi, since specific or even racial specialization of a pathogen provides for the possibility to search for varieties and forms resistant to this pathogen. Vavilov was sure that immune species should be searched for in their country of origin. Of great significance was Vavilov's theory of group or complex immunity, as the breeders were facing the problem of breeding varieties with resistance to a whole population of physiological races, not to a single race. He proposed a scale of resistance to rust strains which has retained its importance until today. An example of natural group immunity is one-grained wheat.

In his last work on immunity, Vavilov emphasized that the detected regularities essentially represented the development of the evolutionary concept as applied to immunological phenomena and thus were leading to an evolutionary or, in a wider sense, genetical theory of natural immunity.

While studying the genetic nature of a host in immunological phenomena, Vavilov found out that diploid wheat species, i.e. one-grained wheat with 14 chromosomes, had high resistance to rust; tetraploid wheats (n=28) were less resistant though they included such species as *Triticum timopheevii* Zhuk. and *T. persicum,* Vav. et Zhuk. which revealed high levels of immunity. And finally, hexaploid bread wheats (n=42) were the most susceptible to rust. The same was discovered by Vavilov for other cultivated plant species such as potato, oat, tomato, sunflower, beet and tobacco.

Taxonomic studies

In August 1930 at the Fifth International Botanical Congress in Cambridge, UK, Vavilov made a presentation of one of his fundamental concepts "Linnaean Species as a System" (Vavilov 1931a). In this report he stated, in particular, the following:

> The Institute of Applied Botany and New Crops (at present the Institute of Plant Industry) has in the last decade accomplished extensive research on the taxonomy and geography of a large number of cultivated plant species according to a definite plan.
>
> The central objectives of this research were the identification of the botanical and agronomic bases of plant breeding, the most comprehensive collecting of the world's basic varieties of the most important cultivated plant species as far as it is possible, and their study as initial material for breeding practice...The research on several hundreds of cultivated species performed by a great number of scientists according to a strictly regulated programme has led us first of all to the understanding of the Linnaean species as a definite complex system, i.e. an integrity consisting of closely interlinked components, where the whole and the parts are merged with each other.
>
> A practical study of several hundreds of species showed the absence of monotypic species, i.e. the species being represented by a certain race alone or a certain form alone. All the species appeared to incorporate a larger or a smaller number of forms (genotypes)... (Vavilov 1965, p. 233).
>
> Detailed examination of the composition of species on the basis of the law of homologous series in variation enabled us to discover a vast diversity of forms and thousands of new varieties and races unknown to a botanist or a plant breeder, many of which are of great practical interest...
>
> Varieties within the species are formed with certain regularity, as it may now be assumed with complete certainty, in contradiction to Linnaeus' views. After obtaining numerous new materials and studying them by all available methods, random phenomena of parallelism in variability observed by Darwin, Naudin and other researchers, have proved to be a general phenomenon, a law, which regulates intraspecific variability... (Vavilov 1965, p. 239).

> Enormous factual materials have led us to a concept of the Linnaean species as a complex system of forms, the composition of which is regulated by the law of homologous series... In the light of modern knowledge of the species as systems of genotype forms, it is necessary to be aware of the system of variability and the amplitude of hereditary differences of separate traits... Existence of regularities in the variability within the scope of the Linnaean species simplifies the study of the systems of their diversity (Vavilov 1965, p. 240).

> The concept of the Linnaean species as a regular system seems highly essential to us not only for a practical study of cultivated plants, but also for analyzing basic problems of the evolutionary process. It is possible to get close to the knowledge of this process only through understanding that the Linnaean species is a complex system, but not the fragments which conventionally serve for description of the species. Solving basic evolutionary problems by a concrete approach cannot be accomplished without taking into account that the species is a complex system of forms (genotypes).
>
> The genetics of separate species gives an idea of the hereditary nature of the species, only when it is based not on several random varieties received from seed companies without any information on their origin, without passport data and, at best, with the name, but on the definitely selected, even occasionally random materials considering the species as a complex system... (Vavilov 1965, p. 242).

> Many Linnaean species represent a complex system of ecotypes or climatypes. An ecotype is a group of biotypes within one Linnaean species united by a series of constant hereditary characters and adapted to certain conditions of their habitat. It is quite natural that differing in space and submitting to the effect of selection, the basic potential of the Linnaean species is to develop and isolate the groups of hereditary forms most closely corresponding to the given environments... (Vavilov 1965, p. 245).

> Thus, the Linnaean species in our view is a flexible isolated complex morphophysiological system linked in its genesis to a certain environment and area of distribution (Vavilov 1965, p. 248).

The origins of cultivated plants

In 1931, Vavilov took part in the work of the Second International Congress on the History of Science and Technology which was held in London, UK. He made a presentation entitled "The problem concerning the origin of agriculture in the light of recent research". (Vavilov 1992). Vavilov decided to draw the attention of the Congress audience to the results of the cultivated crop research activities performed by the Institute of Plant Industry in the last decade. Vavilov said:

> While working with practical problems connected with the breeding of cultivated plants, we approached the solution of a number of problems concerning the history of cultivated plants, applicable to the programme of present research...
>
> It became evident to us, so far, that neither botanists nor agronomists or plant breeders have, in essence, touched upon the inexhaustible wealth of the basic, worldwide resources of even those cultivated plants, the potentials of which exist – as demonstrated by irrefutable research – mainly in ancient agricultural lands. All the plant breeding and the European and American agricultural crops are based on fragments of

the racial composition of such cultivated plants which are derived from the ancient centres of agriculture... (Vavilov 1965, p. 144).

As a result of our studies of some hundreds of cultivated species, we were led to the establishment of basic universal centres of the most important cultivated plants. As presented by us, facts of exceptional general interest were thus revealed.

On the whole, our investigations have led to the establishment of seven basic and independent centres of origin of cultivated plants in the world and, at the same time, to recognize seven definite centres of independently originating agricultural civilizations... (Vavilov 1965 p. 145).

Geographical localization of the primary centres of agriculture is very peculiar. All the seven centres are found mainly in mountains within tropical and subtropical areas. The centres in the New World are associated with the tropical parts of the Andes, those in the Old World with the Himalayas, the Hindukush, the mountains of eastern Africa, the mountainous areas of the Mediterranean countries and the mountains of China, although mainly within the foothills.

In essence, only a narrow belt of the world's continents plays a major role in the history of global agriculture.

Dialectically and in the light of recent research, we have developed a hypothesis of a geographical concentration of the most primitive agricultural crops within these limited areas. The Tropics and Subtropics present optimum conditions for the development of the species-forming processes. The maximum racial diversity in the world, with respect to wild vegetation and wild animals, is concentrated towards the Tropics... (Vavilov 1965, p. 148).

Mountain-building processes no doubt played an important role in the differentiation of plants into species, favouring the development of speciation processes. Factors such as isolation and the origin of barriers against the dispersal of species and genera have certainly always been essential for the evolution of different forms and species. Variation in climate and soil, such as those typical of mountain areas, where the main centres of origin of cultivated plants are concentrated, also coincide with the appearance of diversity among these plants and agree also with the racial composition of cultivated plants...

Although it is mainly a woody flora that has developed in the wet Tropics, herbaceous species have, on the other hand, developed particularly in those mountains of the Tropics and Subtropics where agricultural crops were initially isolated. To the latter belong the majority of the plants cultivated in our world.

The mountainous tropical and subtropical areas present optimum conditions for humans to settle in. Primitive man feared, and to this day fights, the wet Tropics with its unmanageable vegetation and tropical diseases, in spite of the fact that the wet Tropics with their fertile soil occupy one third of the word's landmass (according to Sapper). Man settled, and is still settling, along the edges of the tropical forests, but the mountain areas of the Tropics and the Subtropics offered more favourable conditions for the original inhabitants in the sense of a warm climate and an abundance of food as well as the possibility of life without the need of clothing. The people in Central America and Mexico, as well as in the mountains of tropical Asia, still use a multitude of wild plants. However, it is always easy to distinguish cultivated plants there from the wild ones that are clearly related to them... (Vavilov 1965, p. 150). Knowledge of the basic function of the universal agricultural centres throws light on the entire history of mankind as well as the history of common crops (Vavilov 1965, p. 151).

The data on cultivated plants and animals and ethnographical data on the lifestyle and traditions of various peoples collected by Vavilov during his journeys and confirmed by historical, geographical, archaeological and other fundamental evidence, allowed him to devise a logical system of views on the origin of agriculture and, ultimately, on the civilization of the Earth.

Plant breeding

All Vavilov's theoretical investigations on the centres of origin of cultivated plants and on taxonomy, genetics and immunity found practical applications. All of them served as practical guidance for plant breeders, and plant introducers. In 1934, Vavilov's work *Plant Breeding as a Science* was published (Vavilov 1951). It was dedicated to a theoretical justification of plant breeding as a branch of science. This work retains its significance today and remains an excellent example of the creative development of Darwin's evolutionary concept.

In this work Vavilov wrote:

> In its essence, breeding is the intrusion of man into the natural development of plant and animal forms; in other words, breeding is evolution guided by human will ... (Vavilov 1960, p. 17). From now on the basis of scientific breeding should be formed of precise botanical and goegraphical data on the varietal potential of species and genera (Vavilov 1960, p. 17).

In his work Vavilov defined seven theoretical sections for plant breeding. The first section of this work must have a botanical and geographical base with application of the differential method of plant studies.

> The second section is the concept of variation, while the third one covers interrelations betwen environments and heredity, an organism and external conditions, and, to be more precise, the concept of the genotype and phenotype... The fourth section is the theory of hybridization, brilliantly worked out on the basis of Mendel's laws and Morgan's theory of hereditability. In this section, plant breeding as a science is closer than ever to genetics...
>
> The next, fifth, section we call the theory of breeding process, which means the developments of principles in the work with different plant types: self-pollinators, cross pollinators, intermediate types, etc. This section should include systemized knowledge of the biology of flowering and fertilization as the most important phases determining all the methodology of breeding... These sections, touched upon in general by botany and biology, require differentiation as regards the demands of plant breeding dealing with various groups of plants...
>
> The sixth section is the concept of plant breeding for specific traits: for chemical content, technological or physiological properties, desease resistance and so on. Naturally enough, this section is most closely linked with physiology, biochemistry, technology, plant pathology and entomology... It is the complex approach to a plant, applying different methods of research, that forms the peculiarity of plant breeding as a science. With this, physiology, biochemistry and technology should be interlinked with breeding not only as the sciences used for plant assessment, but even more so to disclose the differences within the most significant crop species and to find out type formation regularities in the most important physiological and chemical properties...(Vavilov 1960, pp. 17-19).

The seventh and last section according to Vavilov should cover individual plant breeding or the concept of breeding of individual plants which is:

> A combination of the knowledge of the individuality of a plant, its differential taxonomy and geography, biology of flowering and fertilization, and the amplitude of variation in all major traits. To gain expertise in a plant, a breeder must perceive his object in its historical and geographical development and find out its differentiation by the most important properties in its interaction with the environments.....This is an updated brief outline, as we understand it, of the essence of plant breeding as a science... Further development of the plant breeding theory might put forward new sections. Advancement of breeding practice will inevitably lead to the development of other biological and agricultural sciences. Embedding definite contents into the above-mentioned sections will require tremendous collective work according to a definite plan (Vavilov 1960, p. 19).

Vavilov came to the following conclusion:

> At the same time, there is no doubt as well that general genetics, backed up by the recorded breeding practice, would conversely receive a mighty incentive for further development. This is witnessed by the history of the evolutionary theory and the history of genetics itself. As a science, plant breeding must, in the nick of time, mount a number of steps and be lifted on an incomparably higher level than the one it is on now. Only further development of plant breeding theory can help a scientist to achieve real control over an organism, i.e. attain the utmost purpose of modern biology (Vavilov 1960, p. 20).

6 Activities associated with plant genetic resources in the 1930s and 1940s

Lysenko's role in the development of agricultural science in the USSR

After the revolution of 1917, Russian agriculture, which had been flourishing, began to decline for physical and organizational reasons. Stalin and his associates faced the challenge of enhancing all spheres of agricultural production in the face of problems due to the food production apportionment practised during Militant Communism, the Civil War of the early 1920s, periodical droughts in the most important crop production areas in the Volga region and the Ukraine, and industrialization with subsequent collectivization of the whole country in the early 1930s. This required a simple and quick solution. At that time, individuals appeared who offered seemingly easy ways of solving the problem. However, they confused what was desirable with what was actually obtainable. One of these individuals was T.D. Lysenko.

Lysenko (1898-1976) was the son of a peasant. He attended horticultural schools in Poltava (1913) and Uman (1917-1920). In 1921 he took courses offered by the Sugar Trust and worked in its experimental stations at Verkhiachka near Kiev (1921) and Belaya Tserkov (1922-1925). He studied agronomy at the Kiev Agricultural Institute from 1921 to 1925. Lysenko first rose to national prominence while posted in Gandzha in Azerbaidzhan (1925-1929). As head of legume selection at an experimental station, he introduced pea varieties from Kiev that were early ripeners, but some became late ripeners in Gandzha. (Soifer 1993). He concluded that this plant characteristic depends less on the breed than on the conditions under which it was grown. He decided that low temperature was related to late ripening and it was possible to produce desirable characteristics by manipulating the growing conditions. In order to make winter wheat sowable in the spring, he suggested that germinating seeds be buried in snow before planting; this reportedly led to greatly increased yields. Lysenko termed this procedure "vernalization" (yarovization – see Chapter 3) and its apparent success brought his work to the attention of agricultural officials. The idea of vernalization, which made Lysenko popular in scientific circles, was formulated by him in1929 at the All Union Congress on Genetics and Plant breeding. In the same year he was given a laboratory in the physiology division of the All-Union Institute of Genetics and Selection in Odessa (Ukraine).

Lysenko's peasant background gave him an important advantage. His techniques may have seemed to offer an immediate way of overcoming the food shortage occasioned by the collectivization of agriculture. Their implementation fitted the social organization of the "kolkhoz" and the "sovkhoz", the collective and state farms which were prominent features of Soviet agriculture (Medvedev 1969).

Work on frost hardiness and the effect of low temperatures had been known long before, and researchers of the Institute of Plant Industry had worked on this subject under the leadership of Dr N.A. Maximov. The Department of Physiology worked on the effects of low temperatures, daylength and moisture on cereal and other crops. However, the impressive and scientifically reliable results promised by Lysenko were never obtained. In his letter addressed to I.G. Eikhfeld at the Polar Experimental Station, Vavilov appreciated and supported Lysenko's work.

> All that is and has been done by T.D. Lysenko is of exceptionally great interest, and the Polar department needs to enhance such activities (Vavilov 1987a, p.134).

At the same time, in his letter to N.V. Kovalev he wrote:

> Lysenko's work is remarkable. It compels us to pose many problems anew. World collections should be worked through in terms of vernalization (Vavilov 1987a, p.174).

In 1932, Yu.A. Yakovlev, People's Commissioner for Agriculture, entrusted Vavilov, as a member of the Board of the Academy, with an errand to review the current state of Lysenko's works and render all kinds of assistance to him. Vavilov wrote to Lysenko:

> ...in August there will be an International Congress of Geneticists and Breeders in Ithaca, USA, and the Commissioner has informed me that if you were willing to attend, the Agricultural Commissariat would apply all efforts to support your trip, so that you could make a presentation of your work there and prepare a display of your activities for the exhibition (Vavilov 1987a, p.165).

When recommending Lysenko for an award in 1933, Vavilov wrote:

> For the first time, with such penetrating profoundness and scope, Lysenko managed to find ways to cope with vegetation control, shift vegetation phases and transform winter crops into spring ones, or late ripening into early. His work is a discovery of primary importance, as it is opening a new sphere for research, and quite an attainable sphere. Undoubtedly Lysenko's work will entail development of the whole branch of plant physiology; this discovery would provide for an opportunity of wide-scale utilization of the world's plant diversity in hybridization for shifting their areas to more remote northern territories. Even the current phase of Lysenko's dicoveries is of paramount interest (Vavilov 1987a, p.188).

Vavilov attributed great importance to this trend in research and always tried to find and identify a grain of truth in this phenomenon, and in the results and practical recommendations made by Lysenko. Vavilov suggested that in agricultural practice this idea could be used to accomplish the shift of southern crops northwards, and in science it would help to regenerate collection accessions of winter crops in the first year of reproduction, or to apply such conditions as a stimulating environment, in selection for breeding purposes and for obtaining hybrids from the parents with asynchronous phases of development.

Vavilov wrote to Lysenko in 1934:

> It seems to me a definite necessity that you, Trofim Denisovich, would yourself spare at least a week, two or three times a year, to come to Leningrad and see what we are doing here, and to help the younger workers especially to perform faster and more effectively the vernalization tasks which are underway here on rather a large scale. You should understand well the significance of such an involvement by you in this work both for us and for yourself (Vavilov 1987a, p.233).

Fig. 37. T.D. Lysenko (in the front wearing a black coat) visiting the Department of Genetics and Breeding at VIR, fields in Detskoye Selo near Leningrad, second half of the 1930s. Second from the right is N.I. Vavilov, the last to the left is I.A. Sizov [VIR Archives].

In 1934, Vavilov, who continued studying this phenomenon, recommended that Lysenko be elected Corresponding Member of the Academy of Sciences. In his reference he wrote:

> Although the nature of vernalization is still a subject of further research and obviously would reveal many unknown aspects, this method has in principle already been developed to such an extent that this year millions of hectares have been allotted for practical vernalization of cereals and cotton. Even now, the great importance of vernalization is evident in plant breeding, enabling the breeder to use all the world's southern diversity which previously could not be grown under our conditions. Moreover, by means of vernalization, many southern varieties are likely to be ready for direct cultivation, even without any breeding, (Vavilov, 1987a, p.219).

During the heyday of Soviet vernalization (1929-1935), Lysenko devised and applied similar techniques to a wide range of vegetables, fruits and grains, and the term itself came to include almost anything done to a crop before planting in order to alter its development to suit local growing conditions, for example, the sprouting of potato tubers before planting. Lysenko and his supporters made extravagant claims for the efficacy of his techniques, and in the early 1930s, vernalization was reportedly applied to many millions of hectares of different crops. Thanks to his work, in 1934, Lysenko became the Scientific Director of the Odessa Institute and a full member of the Ukrainian Academy of Science (Soifer 1993).

Collision of scientific concepts

However, the attraction of the vernalization method soon diminished when the results of its wide application in breeding practice became known. If did not result in the increase in yield which had been widely predicted in the claims of its advocates. Indeed, the discussions grew more and more fierce.

Vavilov continued to search for a rational basis for the arguments of his adversaries. He did this for the sole reason of ensuring that this passionate debate did not lead to loss of whatever could be used for the sake of Soviet genetics and agriculture.

However, it became more and more difficult to do this, because of the discussions covering all aspects of biological and agricultural science in which Vavilov was forced to participate by his opponents. They rejected genetics as a science, rejected the material nature of the gene itself, and rejected Mendel's laws of inheritance. They postulated the effects of environment and various treatments on the heredity of an organism, and the inheritance of only the acquired changes in subsequent generations. VIR became involved in the political processes that were starting all over the country. From the early 1930s Vavilov's scientific programmes were obtaining less and less government support.

At the same time, Vavilov wrote to E.P. Voronov about the start of loyalty checks among the Institute's employees:

> In the near future we may expect a purge. It is my duty to emphasize that all our institutions are definitely loyal to the Soviet State and I hope that we have no serious faults in this regard. The problem of training younger colleagues is evidently quite soluble, since we are a young institution ourselves (Vavilov 1987a, 53-54).

From the early 1930s the poor harvests and the lack of clear governmental policies for improving agricultural production gave Vavilov's adversaries an opportunity to present his work in a deliberately distorted light. At that time the negative attitude toward Vavilov already outlined here became more widespread. Arrests of the members of the Institute's staff started in 1932. Among those arrested were G.A. Levitsky (later he was set free), N.A. Maximov, V.E. Pisarev, M.G. Povov, N.N. Kuleshov and others (Levitskaya 1992; Soifer 1993). In 1934, many leading scientists of VIR left the Institute. Among them were the physiologists I.V. Krasovskaya and V.I. Razumov, the geneticist G.D. Karpechenko, the breeder V.V. Talanov and others (Vavilov 1987a). In this year, it was forbidden to celebrate the 40th anniversary of VIR in spite of official permission having already been given.

In 1935 Vavilov was invited to attend the celebration of the 300th anniversary of the famous Museum of Natural History in Paris. In the same year he was also invited to scientific congresses in Holland and Italy. In 1936, he was invited to go to Czechoslovakia to read a lecture at the Higher Agricultural School in Brno, soon after his election as Doctor Emeritus there. During these years Vavilov made repeated applications to the relevant authorities concerning travel abroad but was not given permission and thus was unable to fulfill his desire to establish scientific contacts. In 1935, the collective monographic work *Theoretical Principles of Plant Breeding* (see Chapter 3) was published. It was his reply to his opponents, but the arguments failed to reach those of his opponents whose efforts were not intended to find the truth. Genetics was declared "reactionary" by the followers of the officially approved science.

In January 1936, Vavilov, along with other scientists, was invited to attend an All-Union Meeting of the foremost agricultural workers held in Moscow. As usual, his presentation was comprehensive and well grounded. Among other reports, one of the most noticeable was the speech of Academician T.D. Lysenko who declared that for the first time he had managed to breed a spring wheat cultivar by planned crossing in an incredibly short time (two and a half years). He assured the audience that, by October 1936, he would be able to release a new cotton cultivar for the southern areas of the Ukraine. However, the promised "successes" did not appear and the techniques applied gave inconsistent results. Nevertheless, these promises were made at the right time to play the role devised by their author.

All the subsequent events developed in such a way that everything proposed by Vavilov's antagonists was approved and supported as being advanced and appropriate for the situation of Soviet agriculture. The research activities and scientific progress of the institutes headed by Vavilov, as well as the results of his personal research, were denigrated or ignored. Thus the conventional principles of discussion in science were violated. Prejudiced assistance and encouragement were given to Lysenko and his followers, and not only agricultural science but also many branches of biological research were brought to the brink of disaster (Soifer 1993).

By early 1936, the standing of Soviet genetics science was high in the international community, as shown by the decision of the International Committee for the Organization of Genetic Congresses to hold the Seventh International Genetics Congress in Leningrad and Moscow. The Congress was scheduled for the second half of August 1937, but it was never held in the Soviet Union. In 1939, it took place in Edinburgh, Scotland. Vavilov was elected President of this Congress. The unprecedented event of a foreign scientist being elected to convene the Congress demonstrated the respect of the genetic community for Vavilov as a first-class geneticist. However, he did not get permission to leave the USSR to travel to England to preside over the Congress.

In December 1936, the Fourth Session of VASKhNIL dedicated to genetic and breeding isssues was held in Moscow. In his report *The Trends of Soviet Plant Breeding* Vavilov emphasized that a prompt solution to the problems faced at that time required a theoretical basis, correct allocation of efforts, planning, coordination and unity in research. He noted that the goal of his presentation was to illuminate the ways that Soviet plant breeding could develop and offer definite measures for enhancemant of its role and importance in socialist production strategy. Then he briefly referred to the history of national plant breeding and to VIR's research on global plant resources. In the conclusion to his report, in dispute with Lysenko on the basic principles of genetics, Vavilov said:

> De Fries was the first to develop the idea of variability of hereditary substance by mutations. Further research, however, did not confirm De Fries'conclusions, and the first decades led experimenters to acknowledge a great measure of stability in genes. This postulate was shaken only by Professor Muller's classical work in 1926 and 1927, where he presented a perfect experimental proof of the possibility of attaining artificial mutations by x-ray techniques...Academician Lysenko put forward a new assumption that the gene is highly variable and may be modified in any way the experimenter sees fit. Precise experimental data have not yet been found to prove it. In the future, Lysenko will probably reveal the experimental possibility of such variation, which will be a new phase that we shall greet when it comes, but since this phase has not yet been confirmed for us geneticists and plant breeders, the lack of an experimental basis for this assumption represents all our difficulties and controversies. Nobody is currently debating gene mutation in genetics as it has been proved, especially in Professor Muller's work and by Morgan's concept, but Muller's explanations and Morgan's discoveries are clearly at variance with Lysenko's statements. No one has yet revealed the exact potential of targetted mutations (Vavilov 1965, p.364).

In conclusion Vavilov said:

....the discussion launched makes geneticists and breeders feel optimistic. We have not convinced each other, but our disagreements have become clear and our viewpoints understandable enough to each other. The first thing needed is more attention to each other's work and more respect for it. We are convinced that in this country, under such exceptional conditions as we are working, when the whole nation keeps watch on our activities, when our achievements are taken over by hundreds of thousands of collective farms, there are all possible prerequisites to fulfil great tasks. Although we disagree on certain theoretical issues, we are pursuing the same goal: we want to make changes in cultivated plants in as short a time as possible, creating the best cultivars in all the most important crops for major agricultural areas. Probably in the near future we will work with different methods taking the best results from each other, but the basic goal will undoubtedly be attained (Vavilov 1965, p.370).

However, hard times followed. First of all, VIR suffered a reduction in its funding for 1937. After the above-mentioned session it became more and more difficult for VIR to publish the world-famous *Works on Applied Botany, Genetics and Plant Breeding*. The Institute's publishing sector was closed and an area in the Stroganov Palace housing a number of scientific departments and the library had to be vacated. VIR came under a real threat of being deprived of one of its basic experimental stations – Otrady Kuban Station in the Northern Caucasus. The situation within the Institute was very complex and tense. Ardent "anti-Vavilovists" were not so numerous among the researchers and doctorate students at VIR, but having unlimited administrative support they were extremely active and lavishly assigned to their opponents the labels which were very popular in the polemics of that time: "Mendelist", "Morganist", "anti-Darwinist" etc., which were later interpreted as having a political significance.

Encouraged from outside, Vavilov's opponents made an effort to create a red-hot atmosphere around Vavilov and the biological research branch of which he was the leader. The late 1930s in VIR were marked by frequent meetings, discussions and sessions where the opposing parties expressed their views. Discussion concentrated on the statement that the effect of environment may produce adequate changes in living organisms. It was said that this is the way "to remodel the nature of organisms in the desirable direction". Very significant was Vavilov's letter to the editors of the *Priroda* journal, dated 22 November 1937, concerning an article written by a G.A. Mashtaler and submitted for publication on the journal. This was entitled "T.D. Lysenko's Concept and Modern Genetics". Vavilov wrote that this article

is unacceptable for publication in "Priroda". It is very contentious, and most of Lysenko's statements are unproven. G.A. Mashtaler often assigns to several cited authors the words invented by himself. He dauntlessly qualifies the modern experimental trend in genetics, including Muller's works, as metaphysical. Such remarks of quasi-geneticists, such as "the environment can affect organisms (genotypes) only in a damaging and destroying manner" do not comply with reality and are simply untrue. It is enough to look through the works of contemporary geneticists such as Muller, Morgan, Dubinin and Timofeyev Resovsky. The fact that modern genetics pays attention to development is obvious because one of the greatest modern geneticists Dr Morgan is at the same time an embryologist. One of his books translated into Russian is entitled "Genetics and Development". A series of basic publications by Morgan have been dedicated to embryology. Ignoring facts and attempting to attribute this or that statement to

> geneticists, the author qualifies as phoenogeneticists or phylogeneticists those who have never been involved in these issues, as for example Burbank.........
>
> The essence of discussion has been understood by the author in a very peculiar sense. Its gist is the fact that several of Lysenko's experimental postulates have been and still are highly doubtful. An experiment may be regarded as proven only if it can be repeated with the same results. As for a number of experimental assumptions made by Lysenko and his followers, unfortunately they demand further and more accurate proof, proceeding from the vast experience of modern genetics. If such proof is submitted, the sharpness of the discussion will be considerably assuaged (Vavilov 1987a, p.360-361).

Prior to the official adoption of VIR's work plan for 1939, in the letter to Karpechenko dated 10 October 1938, Vavilov already urged him to publish an article

> which would manifest what genetics could do. From passivity we need to move to activity. There is no other way out. In the present situation we can oppose profanity with the most convincing facts which are already numerous (Vavilov 1987a, p.385).

Being resolute in protesting against the publication of the submitted article in *Priroda* Vavilov informed the editors that

> if more detailed arguments are needed I am ready to present them (Vavilov 1987a, p.361).

The introduction of Lysenko's ideas and his followers into agricultural practice in the USSR had a serious and deleterious effect. By that time Vavilov's attitude toward Lysenko, and even more so toward his followers, had abruptly changed. In his letter to Yu.Ya. Kerkis in 1937 he wrote:

> I have read your article about Nilsson-Ehle's view upon evolution. In my opinon, it should not be published. For scribblers like Prezent[5] it can give grounds for a rebound, and since his wording is stronger than yours I'm afraid that in the end you will be the loser (Vavilov 1987a, p.359).

There is evidence that Vavilov's friends reproached him for superfluous delicacy and politeness toward Lysenko, saying that by supporting Lysenko he had contributed to his promotion. However, this was by no means a desire by Vavilov to smooth over the sharpness of the dispute at all costs, but an attempt to find every possible way to accelerate the solution of the great practical problems which faced Soviet agriculture.

When Vavilov saw for himself that his ideological opponents appeared unable to contribute anything to achieving the goal of improving agriculture and, moreover, were setting obstacles on the way to it, he turned into an uncompromising fighter against pseudoscience. He rejected the pretensions of Lysenko and his followers who claimed a monopoly in science and the establishment of a specific agricultural biology.

Five lectures on the history of genetics read by Vavilov in Moscow in November and December of 1938 were the obvious proof of his consistency in defending his scientific viewpoint at the highest point of the dispute. These lectures addressed to the audience of doctoral students and younger researchers were the last attempt to keep the young scientific community from groundless discussions and senseless experiments. The lectures provided an extensive and detailed review of references on the history of genetics, and

---

5 Isai Izrailevich Prezent, commentator and intrepreter of Lysenko's ideas.

thoroughly studied the issues covered during the discussion by introducing much factual evidence concerning the interrelations of genetics, breeding practice and evolutionary concepts (Levina 1995).

On 23 May 1939, the extended session of the VASKhNIL Presidium listened to Vavilov's report on the work of the All-Union Institute of Plant Industry in 1938. Despite the enormous amount of scientific research performed by the staff of this institution Lysenko, who was at that time the President of VASKhNIL, did not give his approval to the report. In his closing words after the discussion on the report, Vavilov emphasized once more that the Institute had accomplished a huge amount of work, compiled three volumes on the theoretical basis of plant breeding and had collected priceless plant materials for breeders, employing very qualified personnel for this work. At the same time, the atmosphere surrounding the process of settling theoretical disagreements was absolutely intolerable and anomalous. "History will witness which of us is right", he said in conclusion. Nevertheless, the resolution of the VASKhNIL Presidium considered the VIR Director's report on the Institute's activities to be unsatisfactory (Vavilov 1987a).

Defending the research objectives and the basic concepts of the scientific institute of which he was the leader, Vavilov kept a vigilant eye on the situation regarding genetic research in the country and, in hard times, always helped those who particularly needed aid. In his letter to E.N. Sinskaya of 19 May 1939 he wrote:

> I am busy rescuing the drowning geneticists from the team of Dubinin and Sveshnikova. I do hope we shall save them. The mood is vigorous and enthusiastic (Vavilov 1987a, p.398).

It is obvious from the same letter that Vavilov's attitude toward Lysenko underwent certain changes. Thus, he stated that:

> he [Lysenko] has neither accepted, nor approved of both the plan and the report on the pretext that he doesn't understand them clearly. It is really difficult for us to understand each other (Vavilov 1987a, p.398).

In October 1939, a discussion on genetics was organized in Moscow by the editors of the journal *Under the Banner of Marxism*. Vavilov's addresses were again and again interrupted in the most unfriendly manner by Lysenko, Prezent and others. In his speech Vavilov said:

> Large difference in views on plant breeding methods and basic problems of genetics in this country have arisen to a greater extent by "mutation processes"... If you refer to the very recent times you will see that, six years ago, present-day critics of genetics were writing in the defence of genetics (Vavilov 1965, p.386).

> In this country this interval was marked by great advances in genetics and in practical breeding... new valuable cultivars appeared in Soviet fields and covered dozens of millions of hectares: the cultivars bred in accordance with genetical theory...There is no crisis [for which he was rebuked by Lysenko's followers]. Just the opposite, there is the flowering, and a large scientific school of researchers has been created, which encompasses all the basic branches of modern genetics including the branch most interesting for philosophers: experimental evolution (Vavilov 1965, p.387).

The first of our crucial disagreements is in understanding hereditary and non-hereditary variability... in the concepts of the genotype and phenotype as they were formulated by Johansen. As evidenced by the history of national and foreign plant breeding, the most prominent achievements have been connected with introducing the concepts of the genotype and phenotype in breeding practice (Vavilov 1965, p.391).

Regarding the problem of the material basis of heredity and of the chromosome theory. I would allow myself to say as a biologist that the chromosome theory has been essentially developed for no less than 80 years. It lies at the basis of embryology. It is based on a tremendous amount of factual data. I doubt that one can name another branch of biological science developed to such a level as the chromosome theory (Vavilov 1965, p.392).

The chromosome theory is also exceptionally important in understanding the segregation of characters in remote hybrids, to comprehend the observations of a researcher while crossing remote species and genera...The third point of our disputes and discords, which have been harsh and fundamental, is our attitude to Mendel's laws, to the phenomena of the inheritance of characters in hybrids (Vavilov 1965, p.393).

It is at least strange to deny Mendel after forty years of verification. This is particularly strange for me, since my duties have obliged me to be well aware of the history of genetics. I had an opportunity to study for a long time in England where I observed the struggle which accompanied the establishment of Mendelism and was a witness of the fierce polemics of its first years (Vavilov 1965, p.394).

Now passing over to the opposite point of view which will certainly be better explained by the opponents themselves, we understand that in this field we are being forced to come back again to what existed 30 or 40 years ago or to even earlier times of Gallet, who thought that fertilizers and fostering could alter genetic nature (Vavilov 1965, p.397).

The specific nature of our disagreement lies also in the fact that under the name of advanced science (as opposed to pseudoscience – genetics) we are called to return to the views that have essentially been experienced and overcome by science, i.e. back to the views of the early or middle 19th century (Vavilov 1965, p.398).

Thus, the chief editors of the journal, "Under the Banner of Marxism" must understand that it is difficult for us, the scientists who value the truth and who have dedicated themselves to science, to reject our views. You would perceive the entire difficulty of the situation because everything we defend is the result of enormous creative work, precise experiments in both Soviet and foreign practice...It is necessary to make the editors an offer to prepare and publish translations of the best comprehensive foreign works on plant breeding and genetics, publication of which has been stopped here in recent years....It is necessary to summon conferences and congresses dedicated to the problems of genetics and breeding, so that they could reveal different viewpoints.

The solution of many debatable issues, in essence, can be achieved only by means of direct experiment. It is an obvious prerequisite to offer full-scale possibilities for experiments, even if from the opposite point of view.

And finally the last thing I feel I must stress as a Soviet scientist is that only verified and completely demonstrable results, accurately tested by scientific experiments, should be introduced into breeding practice. In order to introduce them in agricultural production, the proposed measures should undergo accurate testing (Vavilov 1965, p.399).

Regretfully, the organized discussion did not provide for reconciliation of the attitudes of the ideological adversaries toward each other and breeding work based on genetical principles was abruptly curtailed in the USSR.

In November 1939, under pressure from his adversaries and being tired of the disputes distracting him from his work, Vavilov published an order to the Institute's staff concerning a revision of the work plan for 1940 and, in particular, about terminating the genetical research activities in the Institute.

> Following the proposal of the Academy President (Lysenko), I hereby request the revision of the workplan for 1940 from the viewpoint of subordinating it to a fuller extent to the objectives pertaining to the Resolution of the XVIII Congress of the All-Union Communist Party (of Bolsheviks). The plan of scientific experiments should be drawn up in such a way that the execution of the research topics assumed shall provide for increasing the yields in the collective farm/state farm fields and productivity of livestock breeding, enhancing agricultural production, raising labour productivity rates and developing science (Vavilov 1987a, p.410).

In such a difficult situation, Vavilov made up his mind to have a meeting with Stalin; this took place on 20 November 1939. According to Vavilov's ally at VIR Dr E.S. Yakushevsky:

> ... he [Vavilov] after several attempts gained his goal and was granted admission to Stalin at 10 p.m. After he arrived there, he spent two hours in the reception room and at long last, after midnight, was let in. Stalin was already strolling to and fro in his office with a pipe in his mouth. Vavilov entered and greeted Stalin: "Good evening, Iosif Vissarionovich", and bowed. (Later I was told that Stalin hated being called by his first name and patronymic, but preferred being addressed as Comrade Stalin). Stalin did not answer the greeting and said: "So you are the Vavilov who fiddles with flowers, leaves, grafts and other botanical nonsense instead of helping agriculture, as is done by Academician Lysenko, Trofim Denisovich". Here you can see what respect was shown in addressing Lysenko in contrast to simply "Vavilov", and he was not even invited to sit down. Stalin walking up and down the office, leaving Nikolai Ivanovich standing. In the first instance, he was taken aback, but then regained his spirit and read him a lecture about the problems the Institute was busy with, the seed collection from all over the world in the process of being assembled, by then about 200,000 economically valuable samples of various crops having been collected, that everything was done for the development of plant science, development of plant breeding and seed production in this country, and told him about training of new workers and about the great efforts made to organize many research institutions in our country.
>
> Vavilov presented all this, and calmed down, but noticed that it was like talking to a stone wall, you are speaking, and the words are bouncing off him like from a stone wall. At last, after an hour of such talk, Stalin rudely cut him short saying: "You are free, Mr Vavilov". Vavilov bowed and left empty-handed.
>
> He was in similar mood when I met him a week later. His hopes were killed, and he decided that no barriers remained for the followers of Lysenko and that there was no way out for genuine science (or, as they called it, "the bourgeois capitalist science") in the Soviet Union" (Brat'ya Vavilovy 1994, p.38).

An extract from one of his last letters may serve as the finale to the discussion between Vavilov and his opponents. It is dated 24 February 1940 where Vavilov felt all the hopelessness of the situation:

> ... Don't include polemics in your publication: nobody can out-talk the "Prezents", they are too numerous, and the less knowledge they are loaded with, the louder they bark (Vavilov 1987a, p.418).

Only in his work, in the business most dear to his heart, Vavilov felt himself as confident and enthusiastic as ever.

Vavilov's arrest, imprisonment and death. Dissmissal of the Institute's leading researchers

In the summer of 1940, being commissioned by the USSR People's Commissariat of Agriculture, Vavilov led an agricultural and botanical collecting team to the Western Regions of the Ukrainian and Byelorussian Soviet Socialist Republics. This last period of Vavilov's life is recollected here by his disciple and follower F.Kh. Bakhteyev, who himself participated in those events.

> N.I. Vavilov spent a few days in Kiev (Ukraine) where he met the leaders of republican agriculture and scientists and visited several research institutions. On the 27th of July, the mission departed to Lvov. The exploration was to pass through Zhitomir, Berdichev, Khlelnik, Letichev, Proskurov, Volochisk, Podvolochisk, Ternopol, Peremyshlyany and Vinniki. On the itinerary, N.I. Vavilov carefully examined the Plantings and made notes. I remember his excitement when he viewed the vast fields under newly-bred wheat cultivars reaching the old border of the Ukraine, far beyond the horizon. His interest became even more lively after this frontier was crossed. Here we were facing wide fields resembling a patchwork quilt: each tract was under a different crop. As a plant breeder, Vavilov was enjoying it and, in spite of his haste, he very often stopped the car to gather samples of rye, wheat, barley and oats.
>
> Vavilov stayed in Lvov until the 1st of August. During this period as usual he was ebulliently active; he visited the Chief of the Regional Farming Department, and the senior agronomist, thoroughly studied the work of the Agricultural Academy in Dublyany, examined experimental fields, talked to students...and made a visit to the University. Vavilov divided the collecting team into three groups, and sent one group to Volyn, Rovno and Ternopol Provinces, the second one to the hill slopes and mountainous areas of Ivano-Frankovsk and Drogobychi Provinces, and ours (Nikolai Ivanovich himself, V.S. Lekhnovich and F.Kh. Bakhteyev) to Northern Bukovina.
>
> In the morning of the 1st of August, our group left Lvov and headed for Chernovtsy via Stanislav, Kolomyia, Kuty, Vizhnitsa and Vashkovtsy. As usual, on the way Vavilov made frequent stops and collected samples of the planted field crops. When we left behind the former border between Poland and Romania (Northern Bukovina) and moved towards Chernovtsy, near the village of Ispas, Nikolai Ivanovich found among the plantings very diverse populations of oats consisting not only of common cultivated oats, but also of some admixtures of the sandy and oriental species.
>
> Meetings on the way with the inhabitants of Northern Bukovina were warm and friendly. In such cases Nikolai Ivanovich always chatted with peasants, obviously arousing sympathy in the persons to whom he was speaking (Bakhteev 1988, p.215-216).

In his last letter to T.K. Lepin of 2 August 1940 Vavilov was very optimistic when describing his impressions of this trip:

Today I am leaving for Bukovina. I have finished with a half of the Western Ukraine. There are a lot of interesting findings. The hybrids of squareheads with Banat forms are curious, as well as oats. In 2 or 3 days I'll be in the Carpathians. We are beginning to understand the philosophy of Central Europe. There is a lot of science here, including cytology. That is first class botany (Vavilov 1987a, p.420).

Bakhteyev goes on with his recollections:

Late at night on the 2nd of August we arrived at Chernovtsy. Next day Vavilov visited the Agricultural Department of the Ukrainian Communist Party (of Bolsheviks) and the District Land Administration. On the 4th of August Vavilov departed along the route from Zastavna to Zvenyache, where he got acquainted with the experimental fields and got a favourable impression of them. Here it was decided to retain an experimental field in Zvenyache as one of the initial seed production farms. The whole day of the 5th of August Vavilov was at the University, meeting lecturers and researchers, visiting museums and botanical gardens and had some sightseeing in the town...

Following the advice of local scientists, for the 6th of August Vavilov planned a trip to the mountainous area of Putila. Many were willing to accompany him in that trip. Following the recommendation of Nikolai Ivanovich, the author was forced to refuse participation in the trip in favour of one of the guests. Early in the morning on the 6th of August, Nikolai Ivanovich and his companions led their way to Putila. I was commissioned to visit the brewery and inquire about the barley cultivars supplied to it.

This day, after spending a lot of time at the brewery I returned at 5:00 p.m. or so to the University students' hostel where we were staying. It was already getting dark when V.S. Lekhnovich and I returned from the canteen. The gatekeeper told us that a little while earlier Professor (N.I.Vavilov) had come back and wanted to enter the hostel, but at the same moment a car had arrived, several men had left the car and invited the Professor to go with them to have an urgent telephone conversation with Moscow. Then Vavilov had asked the gatekeeper to pass his rucksack over to us and left a message requesting us to wait for him as he would soon return. To our great regret this day we saw Nikolai Ivanovich Vavilov for the last time. As it became clear later, Vavilov had been arrested (Bakhteev 1988, p.217-218).

After his arrest, Vavilov was transferred to Moscow and put into the inner prison of the People's Commissariat of Internal Affairs (NKVD). A criminal case, No.1500, was initiated against Vavilov. The case was processed by Officer A.G. Hvat, an NKVD investigator. In the 1950s, when familiarizing himself with the case of Vavilov in the NKVD Archives, Mark Popovsky found out that a latent investigation was started in 1931, and by the time of Vavilov's arrest the file had grown to 7 volumes. The number of denunciations increased especially after 1937 when relations broke off between Vavilov and Lysenko. This means that Vavilov's arrest was not accidental, but planned by the NKVD. As Popovsky testifies, in the arrest order itself it was written:

It has been established that pursuing the objective of disproving new theories by Soviet scientists Lysenko and Michurin in the sphere of yarovization and genetics, a series of VIR's departments have been assigned by Vavilov with the task of carrying out special work in order to discredit the theories devised by Lysenko and Michurin... (Popovsky 1990, p.191) (English edition, p.146).

Vavilov was charged with sabotage and spying against the USSR. In his book *The Case of Academician Vavilov*, Mark Popovsky presents quotations from Vavilov's criminal file:

> During the first days after his arrest, Vavilov was determined to prove his innocence; his replies at the interrogations were firm and even harsh: "I declare categorically that I have never been involved in spying...
>
> I consider the materials available to the investigation to provide a narrow, incorrect interpretation of my work. Obviously, they may be a result of my disagreement on scientific and managerial issues with a series of persons..." (Popovsky 1990, p.177).

After several days full of strenuous interrogation lasting for 10-13 hours on end, mostly at night, Vavilov admitted himself guilty.

> On August 24, after a 12 hour-long interrogation, the investigator for the first time heard from his victim the words of confession. "I admit, that since 1930 I have been a member of a right-wing anti-Soviet organization existing in the People's Commissariat of Agriculture of the USSR..."- states the protocol of interrogation (Popovsky 1990, p.177).

Only the accusations of spying were completely denied by Vavilov. Yes, he admitted, he had been abroad, visited foreign embassies and missions, but he had never been recruited and never carried out any assignments for foreign intelligence services.

From September 1940 until March 1941, Vavilov was not subjected to further interrogations. In order not to waste time while in solitary confinement, he decided to start writing a book that would sum up his thoughts regarding the global evolution of agriculture since the most ancient times. As Popovsky indicates:

> We know very little about this literary work. Only once did Vavilov mention it in a letter to Beria: "While in the internal prison of the NKVD, when the investigation was conducted and I was permitted to get paper and a pencil, I wrote a voluminous book entitled "History of the development of agriculture" (World agricultural resources and utilization) in which major emphasis is laid on the USSR" (Popovsky 1990, p.192).

In March 1941, the interrogation of Vavilov resumed and charges of organizing an anti-Soviet organization were brought against him and against L.I. Govorov, Karpechenko and other scientists who had been arrested by that time. On 5 July 1941, Officer Hvat completed the investigation of the case which took 11 months and 400 interrogations of Vavilov, mostly at night. A closed session of the Military Collegia of the Supreme Court of the USSR was held on 9 July 1941. Vavilov described this session as follows in his letter to Beria:

> At the trial that lasted several minutes in the conditions of war, I have categorically declared that the charges were based on untrue stories, false facts and slander by no means proved by the investigation (Popovsky 1990, p.200).

Nevertheless, the verdict was as follows:

> The preliminary and full-scale investigations have determined that since 1925 Vavilov has been one of the leaders of an anti-Soviet organization "Peasants' Party of Labour", and since 1930 has actively participated in a right-wing anti-Soviet organization that functioned within the Peoples' Commissariat of Agriculture of the USSR and several scientific institutions of the

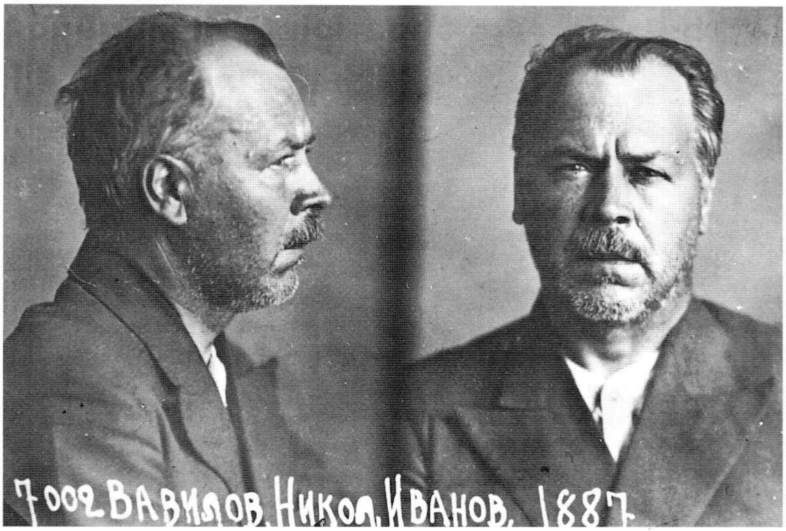

Fig. 38. Photo of N.I. Vavilov taken in prison, August 1940 [VIR Archives].

Fig. 39. Prison in Saratov (modern photograph) [VIR Archives].

Fig. 40. The cell where N.I. Vavilov was placed (modern photograph) [VIR Archives].

USSR... Pursuing the interests of anti-Soviet organizations, he performed acts of sabotage on a wide scale aimed at the undermining and liquidation of the collective-farm system, and the breakdown and decline of Socialist agriculture in the USSR... Pursuing anti-Soviet aims, he maintained contacts with White Army emigrant circles abroad and supplied them with information representing State secrets of the Soviet Union... The Military Collegia of the

> Supreme Court of the USSR has resolved to sentence Vavilov Nikolai Ivanovich to the extreme penalty through execution by a firing squad, accompanied by confiscation of personal property. The sentence is final, and is not open to appeal (Popovsky 1990, p.200).

Only one governmental body, The Presidium of the Supreme Soviet of the USSR, could decide on a suspension of the execution, and Vavilov directed his appeal to it. He wrote:

> I address The Presidium of the Supreme Soviet with a supplication to grant me a pardon and give me a chance to make amends by means of work for my crime against the Soviet Power and Soviet people. Having devoted 30 years to research in the field of plant industry (awarded the Lenin Prize, etc.), I plead for the last chance to complete my work for the benefit of the Socialist agriculture of my Motherland. As an experienced teacher, I swear to dedicate myself entirely to the objective of training Soviet cadres. I am 53 years old. 20.00 P.M. 9.7.1941. Convict N. Vavilov, former academician, Dr Biol. & Agr. Sci. (Popovsky 1990, p.201).

Vavilov waited for a response to his appeal for 17 days. Only on 26 June it became known that Presidium of the Supreme Soviet of the USSR had denied him a pardon. The convict was transferred to the Butyrskaya prison for execution of the sentence. In August, after the beginning of the war between Germany and the USSR, Vavilov directed an application to Beria:

> "In relation to your petition for granting me a pardon and cancelling the sentence given by the Military Collegia, as well as taking into consideration the highest requirements of all the citizens of the Soviet Union in view of the state of war, I dare apply to you for the possibility to concentrate on the tasks in the sphere of my knowledge, plant industry, which are most urgent at the present time:
>
> 1) In half a year, I could complete the "Manual on practical breeding varieties of cultivated plants resistant to major diseases".
>
> 2) In 6 to 8 months of strenuous work, I could complete the "Manual on practical breeding of cereal grasses" applicable to conditions of different regions of the USSR.
>
> Also, I am familiar with subtropical plant production, including such crops of importance for the defence as tung, quinine, etc., as well as plants rich in vitamins. I am eager to place all my experience in the sphere of plant industry, all my knowledge and abilities, in full, at the disposal of the Soviet Power and my Motherland to use them for such needs as I can best serve" (Popovsky 1990, p.222-223).

In mid-October, they started evacuating all prisoners from Moscow prisons to other cities due to the advance of German troops on Moscow. On 24 October 1941 Vavilov, along with a group of other prisoners, was delivered to the Saratov city prison. Eyewitnesses, whom Popovsky has interviewed, recalled that Vavilov died in the prison hospital in January 1943 and was buried in a common grave at a cemetery in Saratov.

The above-mentioned L.I. Govorov, G.D. Karpechenko, as well as N.V. Kovalyov, G.A. Levitsky, A.I. Maltsev and K.A. Flyaksberger were arrested nearly simultaneously with Vavilov. All of them, except for Kovalyov and Maltsev, perished in prison.

After Vavilov's arrest, the leading researchers of the Institute were either arrested, or dismissed and refused permission to work at the leading institutes of the country. Many of them were forced to seek employment with peripheral institutions, but Vavilov's name was never officially mentioned until the middle of the 1950s.

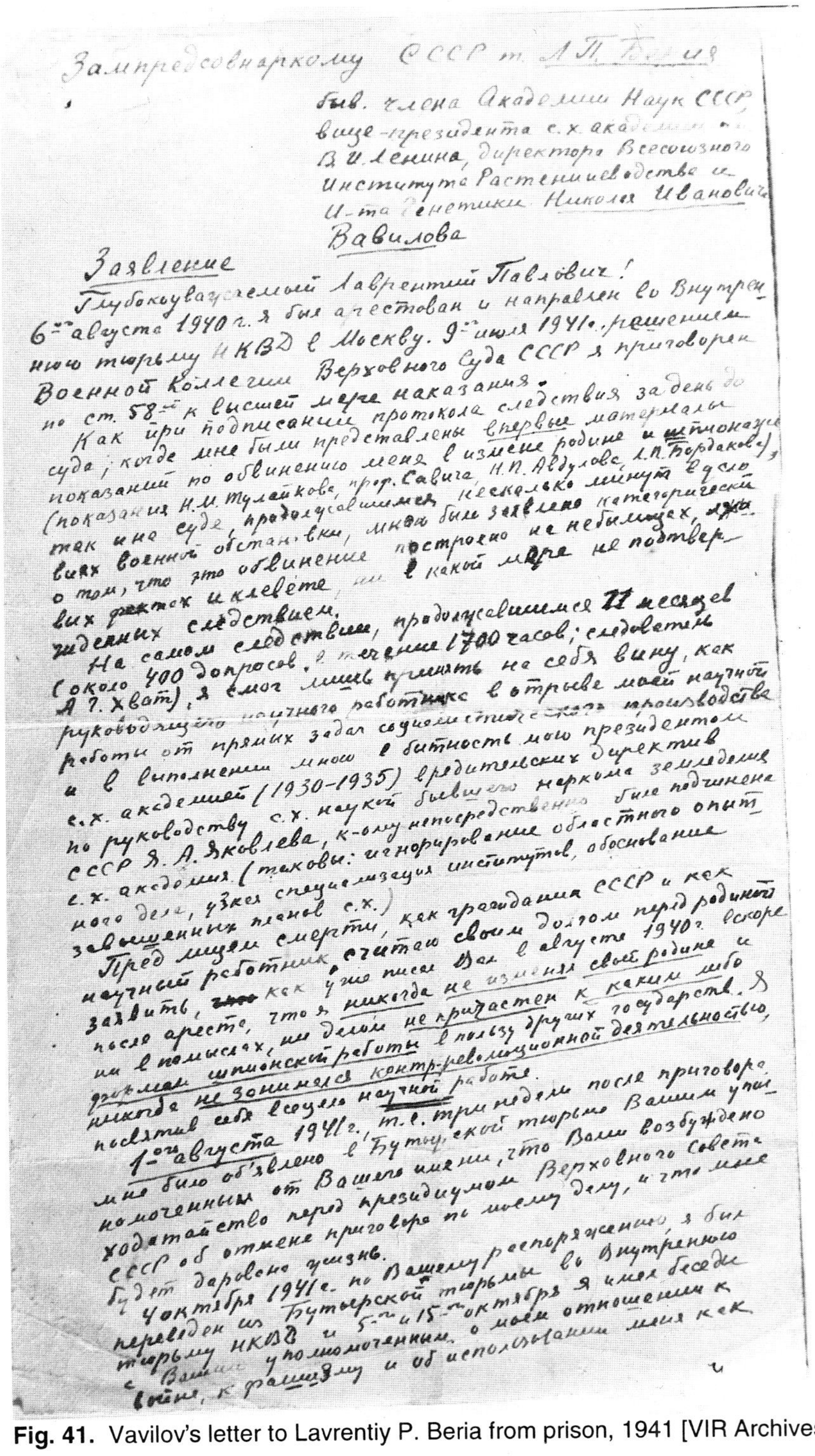

Зампредсовнаркому СССР т. Л.П. Берия

быв. члена Академии Наук СССР, вице-президента с.х. академии им. В.И. Ленина, директора Всесоюзного Института Растениеводства и И-та Генетики Николая Ивановича Вавилова

Заявление

Глубокоуважаемый Лаврентий Павлович!

6-го августа 1940 г. я был арестован и направлен во Внутреннюю тюрьму НКВД в Москву. 9-го июля 1941 г. решением Военной Коллегии Верховного Суда СССР я приговорен по ст. 58-ой к высшей мере наказания.

Как при подписании протокола следствия за день до суда, когда мне были представлены впервые материалы показаний по обвинению меня в измене родине и шпионаже (показания Н.М. Тулайкова, проф. Савича, Н.П. Авдулова, Л.П. Бордакова), так и на суде, продолжавшемся несколько минут в условиях военной обстановки, мною было заявлено категорически о том, что это обвинение построено на небылицах, лживых фактах и клевете, ни в какой мере не подтвержденных следствием.

На самом следствии, продолжавшемся 11 месяцев (около 400 допросов, в течение 1700 часов; следователь А.Г. Хват), я смог лишь признать на себя вину, как руководящего научного работника в отрыве моей научной работы от прямых задач социалистического производства и в выполнении мною в бытность мою президентом с.х. академии (1930–1935) вредительских директив по руководству с.х. наукой бывшего наркома земледелия СССР Я.А. Яковлева, к-ому непосредственно была подчинена с.х. академия. (таковы: игнорирование институтов, обоснование узкой специализации опытного дела, завышенных планов с.х.)

Пред лицом смерти, как гражданин СССР и как научный работник, считаю своим долгом перед родиной заявить, как уже писал Вам в августе 1940 г. вскоре после ареста, что я никогда не изменял своей родине и никогда ни в помыслах, ни делом не причастен к каким либо формам шпионской работы в пользу других государств. Я никогда не занимался контрреволюционной деятельностью, посвятив себя всецело научной работе.

1-го августа 1941 г., т.е. три недели после приговора, мне было объявлено в Бутырской тюрьме Вашим уполномоченным от Вашего имени, что Вами возбуждено ходатайство перед президиумом Верховного Совета СССР об отмене приговора по моему делу, и что мне будет дарована жизнь.

4 октября 1941 г. по Вашему распоряжению, я был переведен из Бутырской тюрьмы во Внутреннюю тюрьму НКВД и 5-го и 15-го октября я имел беседы с Вашим уполномоченным о моем отношении к войне, к фашизму и об использовании меня как

Fig. 41. Vavilov's letter to Lavrentiy P. Beria from prison, 1941 [VIR Archives].

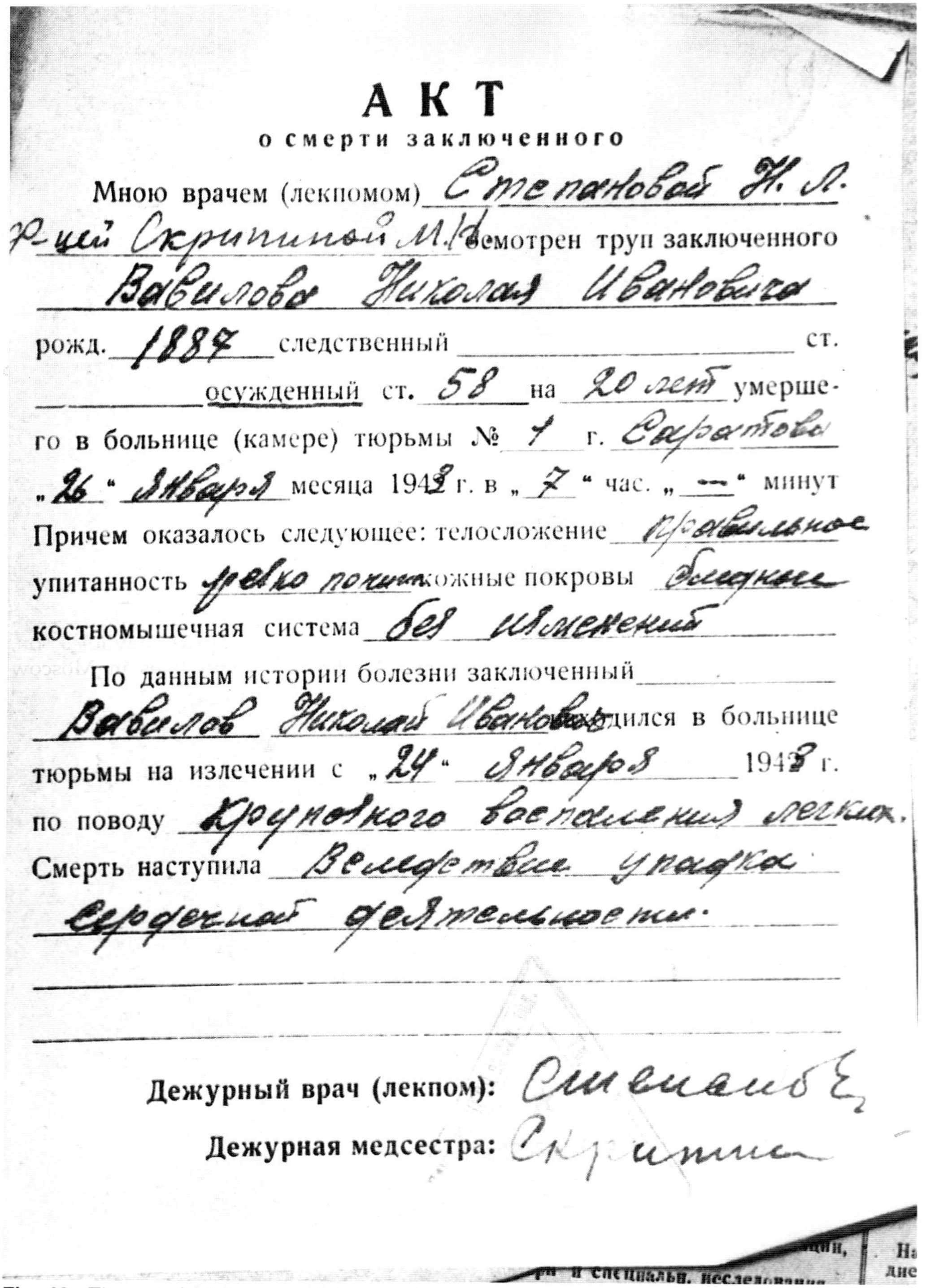

АКТ

о смерти заключенного

Мною врачем (лекпомом) Степановой Н. Л.
Р-цей Скрипиной М. Н. осмотрен труп заключенного
Вавилова Николая Ивановича
рожд. 1887 следственный ________ ст.
________ осужденный ст. 58 на 20 лет умерше-
го в больнице (камере) тюрьмы № 1 г. Саратова
„26“ января месяца 1943 г. в „7“ час. „—“ минут
Причем оказалось следующее: телосложение правильное
упитанность резко пониж. кожные покровы бледные
костномышечная система без изменений

По данным истории болезни заключенный
Вавилов Николай Иванович находился в больнице
тюрьмы на излечении с „24“ января 1943 г.
по поводу крупозного воспаления легких.
Смерть наступила вследствие упадка
сердечной деятельности.

Дежурный врач (лекпом): Степанова

Дежурная медсестра: Скрипина

Fig. 42. The act of 24 January 1943 certifying N.I. Vavilov's death [VIR Archives].

Up to 1941, more then 36 leading researchers of the Institute were dismissed, among them M.A. Rozanova, N.A. Bazilevskaya, E.A. Stoletova, O.K. Fortunatova and F.Kh. Bakhteev. P.M. Zhukovsky left the Institute and took a professorship at the Timiryazev Agricultural Academy (Moscow), where he worked until 1951. N.V. Kovalev and A.I. Maltsev were sent away from Leningrad after their arrest. The latter was sent to Maikop Experimental Station of VIR where he would remain employed until the end of his life in 1948.

Kovalev was exiled to Kazakhstan and worked there as an agronomist at a collective farm until 1946. After that he was transferred to the Middle Asian Branch of VIR (Tashkent) where he was appointed head of the fruit department, and then to the Maikop Experimental Station of VIR. After being dismissed from the Institute in 1949, N.A. Bazilevskaya was employed by the Moscow State University where she would remain for the rest of her life. F.Kh. Bakhetyev was transferred to Murmansk in 1940, then to Moscow in 1943, and in 1949 started work at the Botanical Institute of the USSR Academy of Sciences (Leningrad). After Vavilov's arrest, Rozanova became a lecturer at Leningrad State University and then from 1944 worked at the Main Botanic Garden of the USSR Academy of Sciences in Moscow. Stoletova left the Institute in 1941 and she later lectured in Ivanovo and Kostroma.

E.I. Barulina (Vavilov's wife) retired in 1939 because of health problems. When Vavilov was arrested in 1940, she moved with her son Yury to Moscow. At the outset of war she was evacuated from Moscow to Saratov. At that time she did not know that her husband was dying in prison in Saratov. After the war she suffered hardships as the wife of a "public enemy". When Stalin died she undertook active efforts to clear her husband's name. Her efforts contributed much to the publication of Vavilov's manuscript *World resources of cereals, grain leguminous crops and flax and their utilization in plant breeding* (1957) and helped to start re-editing his other works. E.I. Barulina died in Moscow in 1957. Vavilov's son, Yury Nikolaevich Vavilov, Doctor of Physics and Mathematics, still lives in Moscow (Soratniki 1994).

World War II. The siege of Leningrad and the rescue of the collections

On 22 June 1941 German Nazi troops crossed the Soviet border and quickly began occupying the territories of the Baltic Republics, Ukraine and Byelorussia. By September they reached Leningrad. The German command had a plan to destroy Leningrad, but stout-hearted defenders of the city forced the enemy to halt at the outskirts. Even before the city was surrounded, the Government issued an order for the evacuation of a number of industrial enterprises and institutes from Leningrad. The All-Union Institute of Plant Industry was on the list, but its evacuation was never accomplished.

While some scientists and technical workers were sent to the front, a great number of specialists worked on constructing fortifications around Leningrad. The small number of workers who remained at VIR began the hurried evacuation of the collection (Osazhdennom 1969).

In late August, the collections situated in Pavlovsk and Pushkin were urgently removed under fire to Leningrad. Among them were the collections of potato, rye and other crops. Many difficulties surfaced, particularly with the collection of potatoes. The main portion of this collection – 6000 varieties – was duplicated at the Pavlovsk experimental station, located 45 km southeast of Leningrad. As the harvest of this crop coincided with combat operations near the city, Pavlovsk was ablaze from enemy bombing and the field containing the potato collection was under fire. Clearly, it was imperative that the unripened tubers be dug up. Scientific workers A.Y. Kameraz and

O.A. Voskresenskaya managed to gather this critical collection in a short time, amassing whole allotments of each variety. To help with the task of removing the collection boxes from the field, Kameraz requested assistance from a military unit. The Red Army soldiers respected the importance and needs of the Institute and helped transport the collection by military trucks to a building in St. Isaac's Square (44 Bolshaya Morskaya). For several days this work continued until the fascists seized Pavlovsk (Krivchenko 1991).

In the first autumn of the siege, the Institute lost more than 30 researchers. Some of them perished in the bombing, others died from starvation or were killed in action. During this autumn until the middle of winter the researchers, mostly women, were busy arranging the collection for evacuation.The plan was to remove the collection in two ways. One part of it was to be carried by evacuated employees as hand luggage, for which purpose they selected from 20 000 samples 100 seeds of each cereal sample and 50-200 seeds from each of the samples of other crops. The other part of the collection was packed in boxes with double walls and totalled 100 000 samples or 5 tonnes of seeds, so that the weight of each sample was 20-50 g. A railway van was assigned for this load, and scheduled to be transported to the town of Krasnoufimsk in the Ural Mountains. For six months or more this van remained on side-tracks and was moved from one place to another owing to heavy bombardment of the railways. Afterwards, when all hope of sending the collection away by train was lost, the van was unloaded and the cargo was brought back to the Institute with the help of a military unit.

In winter, the Institute's Director Dr I.G. Eikhfeld, who succeeded Vavilov in this position after his arrest, and several staff members were evacuated to Krasnoufimsk, bringing with them a small part of the collection. Another portion of approximately 40 000 packages of seeds was carried to the same place by aircraft (Osazhdennom 1969).

The largest and most important part of the collection was left in the besieged city. The remaining staff endured the most severe conditions imaginable under the siege during the winter without any heating. In this cruel winter the daily subsistence ration consisted of only 250 g of bread and bran mixture for one food coupon.

In the dark, freezing building of the Institute, the remaining workers prepared the seeds for long-term preservation in the city. While they divided the collection into several duplicate parts, bombs and shells continued bursting around the Institute, damaging St. Isaac's Cathedral nearby. Fortunately, VIR's safety was assured because it was located near the German consulate and the Astoria Hotel where Hitler had planned to hold a victory banquet and had even ordered the printing of the guest invitations. In spite of these difficult conditions, the scientific activities in the Institute continued uninterrupted. A scientific plan was agreed upon in 1942 for the transfer of the most important food and industrial crops to the East. The Institute continued its work despite the unusually severe winter of 1941-42. January and February were the most terrible months of the blockade, with temperatures falling to record lows of minus 36-40°C.

In the winter of 1942, hordes of mice and rats swarmed into the Institute's building at 44 Herzen Street. All efforts to protect the collection against the rodents were fruitless. Rats started invading the metal boxes through existing openings and devoured seeds and grains. Then the malnourished workers decided to tie the boxes together and place them between the shelves. In this way, 18 rooms were used to maintain the collection. Tying up the boxes and their reinstallation proceeded in the freezing premises of the Institute, illuminated only by kerosene lanterns. Shelling had smashed the window glass and the windows were blocked with plywood, which provided more protection for the collection accessions. All the rooms were sealed, and every day the workers checked the seals, also making monthly inspections inside each room. Every day, three to five employees were on duty for 24 hours.

In the spring of 1942 there were several cases of theft of seeds from the guarded facilities. Thieves managed to enter the rooms by crashing through the blocked windows, but the losses after such incidents were not very heavy. The windows were blocked again, while the seeds were transferred to a more reliable place (Osazhdennom 1969).

At this point starvation was raging in the blockaded city, killing dozens of thousands of citizens, among them many of the Institute's workers. In the first months of the siege, E.V. Vulf, a botanist and well-known expert in volatile oil plants, was killed by a shell splinter. In January, A.G. Shchukin, a specialist in groundnuts, died at his writing table. G.K. Kreier, head of the herb laboratory, and D.S. Ivanov, a rice specialist, also succumbed. After Ivanov's death, workers found several thousand packs of rice in his collection that he had preserved while dying of starvation. L.M. Rodina, a keeper of the oats collection, suffered the same fate. Other workers – M. Shcheglov, G. Kovalevsky, N. Leontjevsky, A. Malygina, A. Korzun and others – died of starvation as well. As they slowly starved, they refused to eat the seeds from any of their collection containers of rice, peas, corn and wheat. They chose torment and death in order to preserve Vavilov's genebank (Krivchenko 1991).

The hardest technical task was to preserve the potato collection. V.S. Lekhnovich, who was the curator of the collection during the siege of Leningrad, recalled those events as follows:

> The task appeared extremely difficult. We had to protect the tubers against rats, frosts and starving people. For more safety I started sealing the basement and locked it with three different locks. I also reinforced the doors with iron. However, this could not avert minor thefts... Twice a day, disregarding great emaciation, I struggled on my way from the house at Nekrasov Street where I lived towards St. Isaac Square where the collection was stored. Each walk from one place to the other took about an hour and a half...
>
> The winter of 1941/42 was exceptionally cold. Frost penetrated into the basement where the collection of tubers was preserved from all sides. The stove needed to be heated each day. I was using every opportunity to find firewood. Once a week the superintendent of VIR supplied me with a bundle of firewood. In any case, the temperature in the basement never fell below zero (Osazhdennom 1969).

In spring it was time to regenerate the potato collection. The preserved samples were planted in a field belonging to a suburban state farm. During all summer and autumn the field was guarded against robbers. This procedure was repeated three times in the three years of the siege. By such efforts the collection was saved and partially regenerated. The replanting and regeneration of cereals began under the siege in 1943. N.N. Ivanov remembered this time as follows:

> The work was performed at the State Farm "Predportovyi" under the heavy shelling of the German guns. In an area of 250 sq.metres we planted about 200 varieties (Osazhdennom 1969).

Some samples from subtropical countries and high-altitude places lost their ability to germinate but, on the whole, the collection remained viable.

In February 1944, a group of workers left Krasnoufimsk in the Ural region for Leningrad. There they took a considerable part of the collection preserved in the central building of the Institute and sent it by post for reproduction. In August 1944, at the height of the war, the Institute workers ordered new high-quality collection material of all varieties from abroad. Thus, little by little, the Institute's creative life was revived.

In 1946, after the war had ended, the Institute worked out an elaborate plan to re-sow and restore the viability of the whole genebank at the experimental and state breeding stations. This programme of restoring and maintaining VIR's collection of world varieties was completely accomplished.

Thus, thanks to the heroic efforts of the scientific and technical personnel at the Institute, the genebank of the Institute was saved from destruction and loss of unique germplasm. This heroism cost the lives of many scientists and caused much suffering and deprivation for those who survived. Yet this most dangerous period for the Institute of Plant Industry was overcome.

7 Developments of Vavilov's theories on plant genetic resources after his death

As a result of World War II, the Institute suffered tremendous losses. Many researchers perished in battles or died during the siege of Leningrad. Renewal of the collection was an acute necessity. The Nazi troops destroyed laboratories at VIR's stations in Pushkin, Pavlovsk and other regions, and mined the experimental fields. Thus, the decade from 1945 to 1955 was full of difficulties associated with restoring the Institute and its stations after the war.

In 1951, Academician P.M. Zhukovsky, a Full Member of VASKhNIL, and a colleague of N.I. Vavilov, became head of the Institute. In 1955, after the rehabilitation of Vavilov and his followers, their ideas were brought back to life and during the period from 1957 to 1965 Vavilov's works were published in five volumes by the Academy of Sciences of the USSR. By that time, his first publications had become a bibliographic rarity.

In 1952, a programme of plantings organized on a geographical basis and initiated by N.I. Vavilov in 1923 was resumed by the Institute. These investigations were monitored by P.M. Zhukovsky personally. At first, only spring crops were selected for this third series of sowings (the two initial ones had been performed by Vavilov), and in 1957 a wide range of various winter crops was included in a fourth series.

The results of the third series significantly enriched and refined the results obtained in the first series, in particular regarding variability of the qualitative protein and oil composition. Along with studies on morphological and biological characters, the fourth series was mostly devoted to studying the fractional and amino acid composition of proteins and refraction properties of oils. In 1958 these data served as the basis for the second edition of the *Biochemistry of cultivated plants*.

New information on centres of origin and diversity

E.V. Vulf, who was killed in the seige of Leningrad, had previously identified 16 floristically-rich areas of the globe. This system was taken as a foundation for studying and systematically classifying herbarium materials accumulated at the Institute. Thanks to the resumption of collecting missions by the Institute and the study of the acquired materials, VIR's scientists developed and augmented Vavilov's concepts of the centres of origin of cultivated plants.

For instance, N.A. Bazilevskaya (1964) identified five additional foci of origin of ornamentals, namely North America, South Africa, Australia, the temperate zone of Europe, and the Canary Islands. P.M. Zhukovsky (1970) proposed that four more centres should be added to the seven identified by Vavilov – the Euro-Siberian, African, Australian and North-American centres. According to N.I. Vavilov, these centres are secondary ones, or in other words, the regions of "borrowings" (introduction), as the majority of cultivated plants in these centres, except for a small number, had been introduced from the seven major (initial) centres identified by Vavilov himself. P.M. Zhukovsky introduced a new notion of "megacentres" corresponding to that of the centres of origin by Vavilov, and another one of "microcentres" to describe the foci of introduction into culture of endemic species (forms) which have a narrow distribution. Over 100 such microcentres were identified by Zhukovsky.

According to Zhukovsky, the Euro-Siberian centre covers mostly the central and northern territories of Europe and Asia which gave rise to many fruit species (apple, pear, sweet and sour cherry), nut-bearing crops (dwarf almond, almond, walnut), small fruits (grapes, mountain ash, raspberry, currant, gooseberry, strawberry, buckthorn, etc.), forages (red and pink clovers, alfalfa), vegetables (table cabbage, purslane, onion, horseradish) and industrial crops (flax, dogbane, apocynum, hemp, hop), as well as peas and buckwheat.

The African centre embraces the whole of Africa, including the Abyssinian focus identified by Vavilov and the South African (Cape) region identified by Vulf. The African centre is the homeland of some endemic wheats (*Triticum durum* Desf., *T. turgidum* L., *T. diccocum* Schrank.); rye (*Secale africanum* Stapf.); oats (*Avena abyssinica* Hochst., *A. vaviloviana* Mord.); barley (*Hordeum aethiopicum* Vav. et Bacht.), as well as sorghum, African millet, African rice, coffee, crotalaria, Bambara groundnut, cowpea, castor-oil plant, watermelon, cotton, common flax, Abyssinian cabbage, date palm and banana.

The Australian centre is an important source of new plant materials for breeding valuable crops, for instance *Gossypium* L. (around 10 species), *Nicotiana* L. (over 20 species), 5 endemic species of *Microtus* Swing which easily cross with *Citrus* L. species, 3 species of *Oryza* L., about 400 species of *Acacia*, some of which produce excellent wood. This centre is the primary centre of *Macadamia ternifolia* F.v. Muell., one of the most valuable nut-bearing plants, and is a secondary centre for subterranean clover, as well as of some species of *Panicum, Brachiaria, Eragrostis, Danthoma, Atriplex, Dioscorea, Rubus*, etc.

The North American centre includes numerous endemic cultivated species of *Vitis, Helianthus, Prunus, Ribes, Rubus, Fragaria, Grossularia, Lupinus, Hordeum, Zizania, Gossypium, Nicotiana, Solanum*, etc. (Zhukovsky 1970).

E.N. Sinskaya (1969) continued N.I. Vavilov's investigations of the geography of cultivated plants and augmented his concept of the centres of origin of cultivated plants. Having collated the data from new monographs by plant experts and data provided by archeological and plant collecting missions, Sinskaya revised ideas on the development of cultivated plants and their spread from the centres of origin. She introduced into historical phytogeography a new notion of "affected areas", and proposed to differentiate five major regions of development of cultivated flora, identifying them as the African, Ancient Mediterranean, East-Asian, South-Asian, and the New World regions. Agriculture in North America developed on the basis of Mexican and Central American crops to which crops from the Old World were added later. In central and northern Europe, on the Russian steppes and in Siberia, agriculture is based primarily on cultivated plants introduced from countries in Asia Minor and around the Mediterranean. Agriculture in the "affected areas" is never very old, although the period of development is not really very short either, as can be judged from the large quantity of plants introduced into cultivation from the less rich, wild flora of these territories (Sinskaya 1969).

Continuing the work of his teacher Vavilov, A.I. Kuptsov identified 10 centres of the world's agriculture as the centres of origin of cultivated plants on the basis of anthropological and ethnographic data. These centres were West Asiatic, Mediterranean, Central Asiatic, Ethiopian, Indian, Indonesian, Mexican, Peruvian, North Chinese and West Sudanese (Kuptsov 1975).

The development and organization of VIR's work

Always dominating the Institute's work was Vavilov's principle of strict geographical grouping of accessions for research, regeneration and viability maintenance in accordance with their origin.

At the end of the 1950s the Institute expanded its activities, and the Moscow Branch (Mikhnevo town, Moscow Region), the Krymsk Experiment and Breeding Station (Krymsk town, Krasnodar Territory) and the Yekaterininsk Station (Tambov Region) were incorporated in VIR's network in 1957, 1957 and 1958 respectively. These stations, along with the Maikop Station (Maikop town), the Kuban Station (Kavkazskaya), the Daghestan Station (Derbent), the Far East Station (Vladivostok), the Ustimovsk Station (Poltava, the Ukraine), the Crimean Pomological Station (Sevastopol, the Ukraine), the Aral Sea Station (Chelkar, Kazakhstan) and the Central Asian Branch of VIR (Tashkent, Uzbekistan) became the bases for maintaining and studying collected accessions (Sazonova *et al.* 1994).

From 1961 until 1965 the directorship of the Institute was held by Professor I.A. Sizov, and from 1965 until 1978 by Academician D.D. Brezhnev. On 27 March 1967 the Institute of Plant Industry was named after N.I. Vavilov.

At the same time, the Director of the Institute, D.D. Brezhnev, resumed the publication of the *Works on applied botany, genetics and breeding*, the *Scientific and technical bulletin of VIR*, the *Reference catalogues* containing results of studies on certain genera and species for many years, and later on, in the 1970s, of *Descriptors Lists*.

The methodologies worked out by Vavilov and his associates for evaluating varieties in geographic experiments and for developing an agro-ecological crop classification (see Chapter 3) were desribed in the publication in 1973 of *Methodological guidance directories for studying the VIR collections*. In the early 1970s, descriptor lists were developed for different crops. When the descriptor lists were issued, the Institute started work on the computerization of the passport data of all of the collections.

Phylogeny and genetics

Vavilov's investigations on the phylogeny and genetics of cultivated plants have been continued. Results of these and other continuing studies of the genetic potential of cultivated crops and their wild relatives have been processed by the VIR's researchers into volumes of the *Cultivated Flora* and monographs devoted to individual crops. In the 1950s the Institute published the following volumes of the *Cultivated flora of the USSR*: *Perennial leguminous grasses* (1950), *Perennial grasses* (1950), *Vegetables* (1958); and more recently *Potatoes* (1971), *Root crops* (1971), *Small grains* (1975), *Allium* (1978), *Pea* (1979), *Maize* (1982), *Cucurbitaceae* (watermelon, pumpkin) (1982), *Pome fruits* (1983), *Cabbage* (1984), *Root crops* (1985), *Leaf Vegetables* (1988), *Perennial grasses* (1993) and *Cucurbitaceae* (cucumber, melon) (1994). Second revised and updated editions on major crops have also been published, namely *Wheat* (1979), *Rye* (1989), *Barley* (1990) and *Oats* (1994).

A number of monographs deserve mention: *Soya* by V.B. Enken (1959), *Cultivated plants and their relatives* by P.M. Zhukovsky (1964), *Peas* by R.H. Makasheva (1973), *Barley* by A.Ya. Trofimovskaya (1972), *Rye* by V.D. Kobylyansky (1982), *Wheats of the world* by V.F. Dorofeev (1976, 2nd edition in 1987), *Genetical bases of potato breeding* by K.Z. Budin (1986), among others. Many of them contain taxonomic descriptions of genera and species of cultivated plants and their wild relatives, updated or revised on the basis of new data on genetics, cytology and molecular biology. The publications of the Institute represent an encyclopedia of crop genetic resources. VIR has sent the above publications to numerous research institutions within the USSR and to many foreign countries. (All copies are currently available in the Institute.)

Long-term conservation

Vavilov attributed special importance to preserving for future generations genetic diversity collected all over the globe. Long-term storage became a necessity because the valuable germplasm tended to lose its genetic integrity owing to frequent resowing of accessions in order to restore their viability. Therefore, in the 1940s the first experiments on long-term preservation were launched at VIR. The results obtained made it possible to determine the optimum duration, temperature and conditions for storing accessions of the world collection.

By 1975, the world collection amounted to 230 000 accessions and represented the widest diversity of plant resources collected practically worldwide, most of which are presumably lost forever in nature due to genetic erosion and human activities (Sazonova *et al.* 1994). These efforts benefited from international scientific advances outside the USSR [see FAO publications edited by Frankel and Bennet (1970), Frankel and Hawkes (1975) and Hawkes (1978)].

In order to preserve this rich diversity, the State Seed Store was constructed in 1976. It was built in the Krasnodar Territory, 170 km from Krasnodar, in the grounds of the Kuban Experiment Station. The Seed Store is a three-storey building with the first floor below ground level. The seed storage chambers are located on the ground and the first floors, 12 for each floor. Each chamber can accommodate from 15 000 to 20 000 accessions depending on the size of the container used.

At present, the temperature inside all chambers is maintained at +4°C. Seeds of soya, beans and kidney beans are more sensitive to dehydration than cereals, while some oil-bearing seeds can be dried down to 4% moisture content. The moisture content recommended for long-term storage at the Kuban Station is 4 to 9% depending on the crop. It is clear now that +5°C is the minimum necessary for preserving seed collections, but –10°C and –20°C are the levels aimed at. The old refrigeration equipment is being replaced with a more advanced installation with improved cooling capacity. It is planned to lower the temperature within the storage rooms down to between 0°C and –2°C. In addition, automatically controlled chambers running at –10°C to –12°C are being installed. The storage space allows for the installation of 10 to 12 chambers of this type, with a total capacity of 80 000 accessions (Zaitzev 1990).

Basic research

Following the major objective of developing Vavilov's concepts, the fundamental research at the Institute was broadened, and methodological work on genetics, physiology, immunity, biochemistry and molecular biology has intensified. Investigations by Academician V.F. Dorofeev (1972), who was head of the Institute until 1987, determined that the focus of type-formation and evolution for *Triticum* was located in the Transcaucasus; Professor R.A. Udachin proved Vavilov's thesis of coincidence in Central Asia of the centres of origin, variety-formation, and specific diversity of *Triticum aestivum* L. and *T. compactum* Host. He also described a new endemic species, *T. petropavlosky* Udacz. et Migusch.

Professor A.I. Ivanov (1980) explored Middle Asia (Uzbekistan, Kirgizstan, Turkmenistan and Tajikistan) and Kazakhstan. He found there large foci of introgressive hybridization in wild alfalfa, determined areas of distribution for species with different ploidy, and described the geography of plant characters of value in breeding. Professor N.K. Lemeshev (1992) identified foci of origin of wilt-resistant long-fibre forms of cotton in the Yucatán penisula and other regions of Mexico. Intensive explorations in Western, Central and Southern parts of Africa resulted in the identification by Professor S.N. Bakhareva (1988) of two independent centres of origin and diversity of cultivated plants and their wild relatives: the West African and Central African centres.

Applying the botanico-geographical methods devised by Vavilov, the Department of Tuber Crops headed by Academician S.M. Bukasov and Academician K.Z. Budin studied the geography and phylogeny of South American potatoes, improved the taxonomy of this important crop, and identified a centre of the maximum specific diversity and centres of the most intensive type-formation, as well as areas of concentration of the diversity of traits valuable in breeding.

Researchers from the Vegetable Crops Department (Boos *et al.* 1990; Sazonova *et al.* 1990) have specified centres of origin and intensive type-formation for cabbage, onion, beet, carrot, radish and melon. Hypotheses concerning the evolution of these crops were made, and their intraspecific taxonomy developed (Cultivated Flora 1971, 1978, 1984, 1985, 1988, 1994).

Fruit experts from the Institute studied the intraspecific polymorphism in separate crops, determined foci of maximum concentration of wild pomegranate, pistachio, almond, stone-fruit crops and grapes, specified altitude limits for fig and sour cherry, determined areas of distribution of kerophilous plants, and carried out other research programmes supporting and augmenting N.I. Vavilov's work (Cultivated Flora 1983; Vitkovsky 1984).

The Law of Homologous Series in Variation and the problem of species diversity formulated by N.I. Vavilov have been reflected in studies on molecular biology by Academician V.G. Konarev. The genetics of intraspecific diversity is a high priority among other aspects of the comprehensive study of plant genetic resources (Konarev *et al.* 1993).

With the aim of summarizing the genetic potential of germplasm, the Institute began to publish a series of volumes on genetics of cultivated plants in the 1980s. Three volumes have now been published, namely *Wheat, barley, rye* (1986), *Maize, small grains, oats* (1988), *Legumes, vegetables, melon crops* (1990).

In order to tackle the problems of origin, evolution, geography, and taxonomy of cultivated plants and their wild relatives, specialized departments and experimental stations were organized within VIR. The departments of plant resources in the Institute are organized on the basis of closely related crops. For example, wheat and *Triticale* are studied in the Department of Wheats; rye, barley, and oats in the Department of Cereals; maize, rice, buckwheat, sorghum, etc. in the Department of Maize and Small Grains; clover, alfalfa, timothy grass and other grasses in the Department of Fodder Crops; cotton, kenaf, hemp, sunflower, rape, and groundnut in the Department of Industrial and Oil Crops; and cabbage, onion, tomatoes, cucumbers, melon, watermelon, etc. in the Department of Vegetables and Melons.

The head of each department of plant genetic resources organizes the comprehensive study of the material coming to his department. Members of each department carry out research on the crops and are responsible for understanding the status of their department's crop in all the countries of the world, including native and foreign varieties, achievements in plant breeding, the value of the initial material for plant breeding, prospects for its use, etc. As Vavilov emphasized, every scientist of VIR ought to "have a global perspective", that is, to know thoroughly his crop on a worldwide scale.

In order to study materials from various countries with similar climatic and soil conditions, the Institute established a network of experimental stations. Duplicate collections are preserved at these stations taking into account the ecogeographical origin of each accession, thereby allowing the most favorable conditions for reproduction. By 1985 the experimental stations were spread from the Kola Peninsula (the polar region) to the Caucasus (the subtropics) and from the European part of the USSR to the Far East. The studies were conventionally divided into two stages: field and laboratory assessment. While assessing material in the field, VIR scientists carry out characterization and evaluation. The morphological description of the plants is done afterwards. At all stages of the work materials are compared with the main cultivars of the region. Ultimately, each accession is given a full description.

Laboratory assessment is carried out in special departments and laboratories devoted to Biochemistry, Molecular Biology, Cell and Tissue Culture, Plant Physiology, Plant Immunity, Genetics, Cytology and Anatomy, Technological Assessment and others. Experts in each of these departments and laboratories examine specific aspects corresponding to their field of research. This allows information to be obtained on the value of the accessions, thereby adding to the data obtained by the specialists from the Departments of Plant Resources. As a result of such studies, specialists from the Departments of Plant Resources and the laboratories developing evaluation methods give a final assessment of each accession concerning the prospects of its use in plant breeding and agriculture, and selecting donors of required traits or combinations (Bakhareva 1990).

The Institute also includes the Herbarium where specimens of cultivated plants and their wild relatives are preserved. The Herbarium collection consists of more than 250 000 specimens. To make the work on the collection more manageable, there is a computer system in the Institute, which accumulates basic Herbarium data. In order to standardize the work on the collection and facilitate its computerization, special descriptors have been developed for the crops, including the international descriptors of the former COMECON countries.

The studies of the genetic diversity performed in the field by the Departments of Plant Resources have been intimately connected with the research conducted by the laboratories of VIR. For example, under the directorship of Professor V.I. Krivchenko from 1987, the Institute has been developing theoretical bases, methodological principles and techniques of fungus disease resistance diagnostics in various crop species, system of assessment of the collection for disease resistance, resistance breeding strategies in different crops, problems of genetic uniformity of crop plantings, and their role in epiphytotics (Krivchenko 1984).

The Institute's scientific staff continues Vavilov's work in studying the genetics of the rich plant biodiversity in the VIR's collections and have succeeded in realizing his principles of the practical utilization of research results. Research on comparative genetics of the major species of cultivated plants and their wild relatives is combined with solving evolutionary, taxonomic and breeding problems, as well as meeting the demands of agricultural production (Merezhko 1994b; Mitrofanova 1994; Rigin 1994). Such an approach made it possible to outline a fundamental pattern for genetic studies of plant germplasm for breeding purposes (Merezhko 1984). It comprised the following phases:

1. establishing collections of plant species according to the range of variation in characters;
2. identifying genotypic differences between plant accessions in the characters studied;
3. studying the genetic control of plant characters and determining the number of potentially useful alleles for plant breeding;
4. identifying the alleles;
5. establishing characterized collections.

Special methods were developed to determine the number of genes controlling minor quantitative characters (Merezhko 1994c). These have helped in discovering new genes of value for breeding, developing methods to search for or create donors of such genes, and transfer them into the genotypes of commercialized cultivars. For the first time, genes for cytoplasmic sterility and a dominant form of dwarfness were discovered by Professor V.D. Kobylyansky in rye, thus initiating new trends in rye and triticale breeding (Kobylyansky 1982, 1994). A model of ecogenetic organization of complex quantitative characters has been formulated for the development of new efficient breeding technologies aimed at achieving higher productivity, resistance and quality (Dragavtzev *et al.* 1995).

Under the leadership of Academician V.G. Konarev, storage protein analysis was refined to solve genetic resources questing. These and other approaches have made possible the identification of the plant genome with genomic analysis of allopolyploid species of cultivated plants and their relatives, evaluation of the genomic relationship between species in polyploid complexes of wheat, potatoes, rice, cereal grasses and cruciferous species, discovery of cultivated plant genome origins to determine the degree of relationship and crossability between cultivated and wild species, and control of genome reconstruction, polygenome composition, and introgression in allopolyploid selection (Konarev 1993).

The Institute's scientists have studied the mechanisms of adaptation to stress for various crops, and have modified and developed efficient screening methods. Following Vavilov's concept of developing initial material for breeding, researchers from the Institute study plant genetic resources for valuable characters. As a result of primary evaluation during 1990 and 1991, around 4000 sources of characters useful in breeding were selected, and genetical studies revealed 77 sources. These include 27 resistant to pathogens of wheat, barley, oats, peas, etc., 9 forms of early maturing (for Russian conditions) of oats, maize and sunflower. Fifty-one new breeding sources have been created. VIR experts and breeders from the breeding centres are always in close contact, working together within the framework of the breeding programmes. Breeders come to VIR to get acquainted with new material obtained for the collection, while VIR researchers go to the breeding centres to study breeders's needs and requests for breeding material. In addition, special field workshops are organized at VIR experimental stations, and breeders are invited to participate. At these workshops breeders also get acquainted with new accessions in the field and decide together with the VIR experts which of the accessions are more promising for breeding programmes.

As a result of such cooperation, more than 60% of commercial varieties have been bred using material from the VIR world collection. In cereals, 95% of all cultivars bred in the recent period were based on germplasm stored in the VIR collection. Using these materials, USSR breeders developed over 2,500 cultivars of various agricultural crops. In the 1990s many of them were commercially cultivated, occupying over 60 million hectares (Sazonova *et al.* 1994). Collections of such important crops as wheat, barley, rye, oats, maize, pea, soya, potato, tomato, cucumber, apple, cherry, plum, strawberry, alfalfa, etc. were used most intensively.

The comprehensive study and use of new initial material made it possible to broaden breeding programmes for the major crops considerably. Materials from the collections served as a basis for developing sunflower cultivars with high oil content, the first national heterotic hybrids of maize, potato cultivars possessing resistance to *Phytophthora*, aggressive biotypes of potato wart disease *Synchytrium* and nematodes, and high-yielding cultivars of cereals and vegetables. A new trend in breeding dwarf wheat and rye was started. Wild cotton species have been used in breeding for leaf-drop in order to exclude pre-harvest chemical defoliation, and to serve as sources of gossypol-free seed. Tomato forms possessing genes necessary for creating cultivars and hybrids with an increased fruit keeping ability have been used. Cytoplasmic male sterile lines of sugarbeet and other vegetable crops have been introduced into breeding programmes.

Selection of appropriate materials made it possible to start breeding rice for semi-dwarfness, peas for leaflessness and non-shattering of pods, and soya for nematode resistance. On the basis of studies of global plant diversity, researchers in the Institute have included new aspects in breeding; for example, enhanced nitrogen fixation ability in leguminous crop cultivars and resistance to mechanical damage in potato varieties. Newly created Russian cultivars of the cereal crop *Triticale* have been created and successfully introduced into cultivation by VIR specialists (Sazonova *et al.* 1994).

Effective use of the world's plant genetic resources combined with success in plant breeding justify the priority given in the Institute's to collecting global plant genetic diversity.

Post-war collecting missions

Russian expeditions

In the post-war period, in addition to the regeneration of older pre-war samples, the Institute was facing the problems of restoring lost material, replenishing the collections with new accessions and exploring new centres of diversity of cultivated plants and their wild relatives. With this purpose, the first post-war collecting missions began in 1946. These initial missions were sent to different regions of the USSR with the aim of collecting local varieties and landraces.

In 1946-1948 explorations were carried out in the Leningrad Region, Estonia, the Pskov Region, the Trans-Carpathian area, Daghestan, Nagorny Karabakh, Kazakhstan and Kirghizia. From the 1950s the Institute continued these collecting trips within the USSR on a regular basis, first of all exploring the territories of Central Asia, Caucasus and the Far East as centres of diversity and origin of cultivated plants and their wild relatives. In 1951, for example, VIR's missions collected 2177 crop samples of cereals, grain legumes, vegetables, fruit and berries (Shcherbakov *et al.* 1971).

By 1965, the Institute had carried out 130 collecting missions in various regions of the Soviet Union. With the objective of wide-scale and systematic exploration of plant genetic resources in the USSR, in 1966 the Institute established a collecting missions planning group: European, Caucasian, Kazakhstanian, Cenral Asian, West Siberian and Far East ones, and since 1976, the East Siberian mission (Sazonova *et al.* 1994).

Collecting missions exploring the European part of the USSR brought winter-hardy and early-ripening forage grasses resistant to prolonged springtime flooding. Siberian expeditions replenished the collection with various species of forage grasses capable of surviving excessive moisture, with intensive spring aftergrowth, drought resistance and salinity resistance. They included interesting samples of early-ripening radish with white delicate flesh, cultivated onion of good keeping quality and wild onion species. A large collection of currants and sea buckthorn was brought from the Ural Mountains. Explorations in Kazakhstan found the most productive plant communities of forage grasses resistant to drought, salinity, frost and seed shattering.

A unique collection of melon landraces and all available forms of apricot and grape were collected in Central Asia. Ancient sites of pomegranate cultivation were also explored in this region. Explorations in the western Kopet Dag enriched the collection with high-yielding parthenocarpic forms of fig and diverse forms of pistachio and almond. The genebank also received samples of naked barley from the Pamir areas, plus forms of *Triticum aestivum* L. and *T. compactum* Host. without ligules.

Almost all of the specific potential of *Triticum* L. was collected in the areas of the Caucasus and Transcaucasus. Of particular interest was *Triticum militinae* Zhuk. et Mig. with complex disease resistance. Great attention was paid to collecting wild and weedy field species of *Avena* L. In this region, collecting missions found samples of *Hordeum bulbosum* L. and rare genotypes of *Secale montanum* Guss. resistant to various diseases. Among the samples collected were wild relic radish (*Raphanus rostratus* DC.) and wild beet species (*Beta corolliflora* Zoss., *B. intermedia* Bunge, *B. macrorrhiza* Stev.). New areas of distribution were identified for wild perennial beet (*B. trigyna* W. et K.). The collection was replenished with numerous and diverse forms of blackthorn, quince, almond and subtropical plants.

Explorations in the Far East resulted in the collecting of interesting wild forms of onion, wide diversity of Ussuri soyabean and various forms of *Phaseolus* beans. They helped to define more precisely the areas of distribution of Chinese magnolia vine. The collection incorporated accessions of Amur barberry, different currant species, Amur grape, *Actinidia* and honeysuckle.

From 1966 to 1993, the Institute's scientists undertook botanical and geographical surveying, and collected plant germplasm from the territories stretching from the Kola Peninsula to the Black Sea Coast of the Caucasus, and from Belorussia to the Kurils. As a result, over 60 000 seed accessions and planting materials of agricultural crops and their wild relatives were added to the Institute's collection, while the herbarium was enlarged by approximately 100 000 specimens (Mesherov *et al.* 1981; Kobylyanskaya *et al.* 1990; Zoteeva *et al.* 1996).

Every year, 30 to 40 collecting teams were engaged in collecting plant genetic resources in the USSR. A collecting trip usually lasted from 15 to 45 days. These missions had narrow specializations, each focusing on collecting species of *Triticum, Aegilops, Hordeum, Linum, Pisum,* fodder crops, vegetables, or fruit trees, etc. As some of the collecting sites were distant and hard to reach, each group was staffed with scientists who were both specialists in various crops and also able to collect all plant materials to be found in the wild. During explorations, each collecting team gathered from 50 to 300 accessions plus, if possible, herbaria samples (Bakhareva 1994).

In post-war times, tension in international relations and other political difficulties made it almost impossible either for foreign scientists to carry out exploration and collecting in the USSR, or for Soviet scientists to conduct collecting missions abroad. Only in the early 1960s were Soviet and American scientists able to establish contact and plan collaborative collecting and study of plant genetic resources under the Soviet-American Joint Commission for Cooperation in Agriculture including the Ministry of Agriculture of the USSR and the USDA.

In the 1960s, three Soviet-American joint collection teams with the participation of Drs J. Creech and D. Scott (1963) travelled to the Crimea (Ukraine) and the Central Asian republics and collected fruits and ornamental plants; Drs W. Keller and Q. Jones (1965) travelled to the Krasnodar and Stavropol regions of Russia and collected forage crops; and Drs W. Skrdla and H. Brooks (1967) explored the territory of the Crimea (Ukraine) and the Stavropol region (Russia) and collected fruit and forage crops.

In the 1970s cooperation with the USDA continued: Dr J. Creech visited the Crimea and Siberia in 1971 to collect forages and ornamental plants; Drs D. Dewey and A. Plummer travelled to the Caucasus and Central Asian republics in 1975 to collect perennial forage crops (in particular *Agropyron, Elymus, Hordeum, Trifolium, Trigonella, Onobrychis, Medicago* and *Astragalus*).

Within the framework of the programme "Collection, Documentation, Preservation and Exchange of Seeds and Other Planting Materials of Agricultural Crops" endorsed by the Soviet-American Joint Working Group, a joint collecting mission was undertaken in 1977 with Drs Dewey and Plummer taking part in it from USDA. The plant collectors explored the territory of the Stavropol Region, Russia, and several areas in Kazakhstan. As a result, more than 1100 accessions of wild perennial forage grasses (*Agropyron, Bromus,* etc.) and legumes were collected.

Within the same joint programme, three collecting missions were carried out in the 1980s with the participation of specialists from the USDA. Drs M. Rumbaugh and K. Asay visited the USSR in 1982 to explore the territories of the Ukraine, the Caucasus regions (Russia) and Uzbekistan to collect forage grasses and legumes. In 1988, Drs K. Asay, D. Johnson and D. Casler took part in joint plant explorations in Siberia, Russia, and Kazakhstan to collect *Dactylis, Agropyron, Elymus, Medicago, Festuca,* etc. In 1989 Drs P. Simon, L. Pike, J. Swenson and T. Kotlinska (Poland) in association with VIR's specialists explored the territories of Kirghizia, Uzbekistan and Tadjikistan to collect genetic diversity of *Allium* spp.

Toward the end of the 1980s, VIR expanded bilateral cooperation with various genebanks and international organizations on collecting cultivated plants and their wild relatives. For such purposes, VIR organized joint missions with foreign scientists to the territory of the former USSR. Most of these explorations were supported financially by the foreign participants.

1989 was the starting point of the joint collecting missions under the aegis of IBPGR.[6] The first collecting trip of VIR specialists plus scientists from Southampton University (UK) and from New Zealand was realized in two phases by two teams (Drs N. Maxted and P. Munylmyembe from UK; Dr F. Bisby from UK and Dr M. Forde from New Zealand) in Armenia, Azerbaijan, Georgia and Russia (Daghestan) to collect *Astragalus, Coronilla, Lathyrus, Lotus, Medicago, Onobrychis, Trifolium, Vicia,* etc. In 1990 there was a joint mission under the auspices of IBPGR for wild legumes (*Lens, Cicer, Pisum, Vicia, Lathyrus* and other leguminous crops) with Dr N. Maxted (UK) and Dr G. Ladizinsky (Israel). In 1991, the third joint trip was launched by IBPGR and VIR, in which Drs N. Maxted (UK) and C. Sperling (ARS/USDA) collected wild legumes (*Trifolium, Medicago, Lathyrus, Lens, Vicia,* etc.) in Tadjikistan, Kirghizia and Uzbekistan. In 1990, the Australian scientists Drs K. Reed and B. Dear undertook joint collecting of pasture grasses in the Crimea (Ukraine), Krasnodar and Stavropol Regions (Russia). In 1990, VIR launched a joint collecting mission for fruit germplasm (*Prunus, Pyrus, Juglans,* etc.) with Drs C. Sperling, D. Ramming and M. Thompson. In 1991 a mission including Drs T. McCoy and C. Sperling explored the Krasnodar and Stavropol regions, Ossetia and Daghestan (Russia) and Uzbekistan for *Medicago* spp. In 1992 Drs K. Asay and D. Johnson went to several regions of Kazakhstan for *Agropyron, Festuca, Poa, Phleum* and other forage plants.

In 1990-91, there were two collecting missions in cooperation with the Centre for Genetic Resources the Netherlands (CGN), along with Dr L. Frese from Germany, to explore the diversity of *Beta* spp. in Armenia, Georgia and Russia (Daghestan). In 1991 Dr M. van Slageren from the International Centre for Agricultural Research in the Dry Areas (ICARDA) participated in a joint ICARDA/VIR mission exploring genetic resources of *Aegilops, Triticum* and *Hordeum* in Uzbekistan and Turkmenia.

In 1991, VIR established close contacts with the Japanese National Institute of Agrobiological Resources (NIAR). Drs H. Daido and H. Shimokoji from NIAR and S. Shuvalov (1991) explored Khabarovsk Province and Sakhalin Island, searching for germplasm of *Festuca, Phleum, Dactylis, Poa,* etc. In 1992, Drs H. Yamaguchi and E. Monma (1992) collected germplasm of *Trifolium, Medicago, Galega, Melilotus, Dactylis, Lolium, Phleum, Festuca,* etc. in the Krasnodar and Stavropol Regions, Ossetia, Daghestan (Russia) and Kazakhstan. In the same year, Drs T. Sanada, H. Besho and S. Shuvalov collected seed and scionwood of *Prunus, Malus, Vitis, Pyrus, Hippophae, Crataegus, Berberis, Cerasus,* etc. in Uzbekistan, Kazakhstan, Kirghizia and Tadjikistan. In 1993 Drs K. Okuno and H. Yoshida went to Turkmenia, Uzbekistan and Kazakhstan to collect the diversity of *Triticum, Aegilops, Hordeum, Avena,* etc., Drs A. Kojima and T. Yosida collected species of *Allium, Cucumis, Citrullus, Cucurbita, Daucus,* etc. in Uzbekistan, Kazakhstan and Kirghizia, and Drs N. Tomooka and J. Arihara explored Kirghizia, Uzbekistan, Tadjikistan and Turkmenia to collect germplasm of *Vigna, Phaseolus, Cicer,* etc.. In 1994, Drs K. Okuno and H. Yoshida collected *Aegilops, Hordeum, Avena, Secale* and other cereals in Krasnodar and Stavropol Regions, Ossetia, the Chechen Republic and Daghestan, and Drs K. Shinoda and O. Urashima

---

[6] IBPGR – International Board Plant Genetic Resources; which became IPGRI – the International Plant Genetic Resources Institute in 1994.

collected genetic resources of *Allium, Tulipa, Fritillaria,* etc. in Uzbekistan and Kazakhstan.

During this period, the germplasm collection of the Vavilov Institute was enlarged annually by 2000-3000 accessions collected within the former USSR. For example, in collecting missions in 1989 involving 136 scientists, VIR explored 59 areas of the country and collected 2100 accessions (see Appendix III).

Overseas collecting

Vavilov's theories of the "Centres of Origin of Cultivated plants" and of the "Law of the Homologous Series in Variation" served as the theoretical basis for all of the plant collecting activities of VIR. These laws provided invaluable information for targeting the research at specific germplasm or at a desirable trait. In choosing plants for replenishment of the existing collection, the following principles were used: to collect botanical forms and varieties still missing from the collection; to introduce sources of valuable breeding characters and donors of particular genes; and to accumulate the maximum phenotypic and genotypic diversity available in the collected species. In keeping with Vavilov's traditions, after each collecting mission or scientific trip to a foreign country, participants made a very detailed report on the situation of plant science, on the level of major crop breeding activities, systems of collecting and maintenance of plant genetic resources of the given country, etc. and published the main results of these missions in the Institute's Works.

1950s

In the mid-1950s VIR was able to resume germplasm-collecting activities in foreign countries. The period allocated for a journey to a foreign country varied, depending on the domestic and foreign policy of the USSR and the international situation in general, relating to the tension between the capitalist and socialist systems.

In the period from 1954 to 1966, plant explorations were conducted for the most part in the countries bordering the Soviet Union (Shcherbakov *et al.* 1970). Scientific trips to West European countries were very brief, although they helped to add a certain number of the newest cultivars and old landraces of different crops to the VIR germplasm collection.

Professor P. M. Zhukovsky made the first post-war scientific journeys to France (1954), Italy (1954, 1960), Argentina (1955, 1958), Chile, Peru and Mexico (1958) where he visited leading national breeding institutions, re-established the potato collection, collected seed samples of cereals, legumes, vegetables and many other crops, and brought back many publications on plant breeding.

In the 1950s, VIR scientists commenced a detailed investigation of the Asian centres of crop origin. They collected new plant materials promising for breeding from China (N.R. Ivanov and V.T. Krasochkin in 1952-1953, 1956; D.I. Tupitsyn in 1956, 1957), India, Nepal (D.V. Ter-Avanesyan in 1956-1959), Mongolia (N.G. Khoroshaylov in 1957-1959) and Iraq (T.N. Shevchuk in 1959). During the same period, collecting missions sent by VIR (1956, 1959-1960) made detailed explorations of the whole territory of Bulgaria with the purpose of collecting local varieties and populations of bread and durum wheat, maize, oats, *Phaseolus*, pea, vetch, vegetables and forage crops. Those efforts added about 2400 samples of all the crops cultivated in Bulgaria to the collection. In 1958-59, the first post-war expedition to Africa took place and the VIR collection was re-supplied with germplasm from Egypt, Ethiopia and Sudan that had been lost in part during the war.

1960s

In the early 1960s Professor D.V. Ter-Avanesyan visited Indonesia (1960) and Japan (1964) in order to get acquainted with the work of breeding institutions of those countries and to collect local and new plant samples. In 1963, an important plant genetic resources collecting mission was sent to Afghanistan. Being the first mission since N.I.Vavilov's trip in 1924, it helped to identify the changes in the specific and varietal composition of this country's crop vegetation and recover lost accessions. With this purpose, the collecting mission explored the main agricultural areas of Afghanistan, studied the distribution of plant genetic resources, and collected over 700 seed samples and planting materials representing local varieties and breeding cultivars of various crops.

From the late 1960s there was a gradual increase in the acquisition of plant genetic resources from abroad. A new germplasm collecting campaign started in 1967, when it became possible to organize regular collecting trips to the centres of diversity of cultivated plants and their wild relatives and visit countries with well-developed breeding programmes. In this period, trips were made to the traditional countries of the Asian and African centres having a wide diversity of cultivated crops: Turkey, 1967; Iran, Ethiopia and Sudan, 1968; Afghanistan, Japan, India and Algeria, 1969; and the first explorations in Tanzania and Uganda, 1968. VIR also made the first post-war exploration and collecting trips in South America (Chile, 1967; Mexico and Brazil, 1968) and Australia (1968) at this time.

The centres of origin described by Vavilov were not only visited by Soviet collecters, missions were also arranged by the USA, Australia, India, Italy, Japan, Germany and other countries. From the time of the first publication of Vavilov's concept of centres of origin until 1970, scientists from the USA carried out 101 collecting missions, 75 of which were in centres of origin. From 1967 to 1976, Australian scientists undertook 58 germplasm collecting trips, 45 of which explored centres of origin (Kobylyanskaya 1989).

1970s

From the middle of the 1970s, when VIR was most active in establishing an international plant genetic resources network, its germplasm-collecting activities became ever wider. Each year, the Institute organized several collecting trips to three or four continents with the goal of exploring centres of crop origin and neighbouring countries. In Europe, there was a thorough exploration of Spain in 1977 for the first time since N.I. Vavilov's mission in 1926. The Institute's scientists visited all areas of plant cultivation and collected over 5000 samples mostly of local cereal, legume, vegetable and fruit crops. Introduction of new cultivars from European countries placed at the disposal of breeders material such as heterotic hybrids of sunflower and maize, rape with a low erucic acid content, and nematode-resistant cultivars of potato and oats, as well as the newest cultivars of vegetable and fruit crops, flood-tolerant forage grasses, lodging- and (late) sprouting-resistant cultivars of rye and wheat.

The Asian centres of diversity and adjacent countries were explored by the Institute's teams in 1970 (Pakistan), 1974 (Iraq, Nepal and Syria), 1977 (India, Korea and Philippines), 1978 (Yemen, Afghanistan and Japan) and 1979 (Bangladesh and Syria).

In the 1970s, great attention was paid to studying the centres of crop diversity in Africa and South America. In the African continent, taking into account the centres of diversity described by P.M. Zhukovsky, the Institute's researchers explored the northern part (Tunisia in 1970; Algeria in 1978, and, for the first time, Libya in 1974), and eastern regions (Egypt in 1974; Sudan in 1978; and, for the first time, Kenya in 1970, Burundi and Somalia in 1972). They also explored west Africa for the first time (Guinea, Mali, Senegal in 1972; Nigeria in 1974; Burkina Faso and Ghana in 1977) and Madagascar (1979).

In South America, major collecting missions were concentrated on the species which had originated in American centre, of diversity such as cultivated and wild species of potato, maize, sunflower, cotton, tomato and other crops. Explorations were carried out in

1970 (Peru), 1971 (Peru, Bolivia and Ecuador), 1973 (Argentina and Uruguay), 1975 (Mexico), 1976 (Colombia and Venezuela) and 1978 (Jamaica, Trinidad and Tobago). VIR's scientists also made several trips to the USA (1977, 1979), Canada (1979) and Australia (1971, 1975, 1977) in order to study breeding practices and to exchange plant germplasm.

Collecting in Australia yielded promising forage grasses capable of rapid growth after cutting, cotton cultivars characterized by uniform boll ripening, resistant forms of wheats and high-quality vegetable crop cultivars (Kobylyanskaya 1989).

1980s

In the 1980s the Institute organized expeditions to all continents, in recognition of Vavilov's instruction that explorations of the same territories, especially in the centres of crop origin where the process of plant type formation was most dynamic, should be performed every 5 or 10 years. In Europe the Institute's scientists concentrated on the Mediterranean centre of origin and diversity of cultivated plants and their wild relatives. They surveyed Yugoslavia (1980), the Canary and Balearic Islands (1982), Italy (1983), Greece, France (1984) and Spain (1986). They also undertook trips to a number of countries in Western Europe where the level of breeding practice was high (Austria, Norway, 1981; Netherlands, 1982; Belgium, 1983) and Eastern Europe (Poland, 1983, 1989; Bulgaria, 1985-1986; Hungary, 1985, 1988), so as to obtain new plant breeding materials and information on modern breeding trends.

In Asia, the first collecting missions were sent to Burma (1980), Laos (1983), the Yemen Arab Republic (1984) and Bhutan (1989). The demand to make the VIR collections more complete led to explorations in North Korea (1980-1981, 1987), Turkey (1982, 1985), India (1983), Sri Lanka (1985), Mongolia, Syria (1986), Nepal (1988) and China (1989). From the Yemen, a unique region in the Arabian Peninsula, drought-resistant forms of wheat, barley, sorghum and African millet were collected, plus extremely early forage grasses and drought-resistant samples of cowpea, lentil and beans. Aboriginal forms of various agricultural crops from Laos and Burma also deserve mention.

During these years, the Institute's researchers continued exploring of new areas in the African continent for the first time, and visited such equatorial and southern African countries as Gabon, Congo (1980), Zaire, Zambia (1982), Benin, Botswana, Zimbabwe (1986) and Côte d'Ivoire (1989). From there, large collections of tropical plant species and genera were sent to the Institute together with samples of traditional crops adapted to Africa and having traits of breeding value such as disease resistance, drought resistance and heat tolerance. Collecting also took place in Burundi, Tanzania, Uganda (1983), Nigeria (1984) and Morocco (1987). In Zaire and Zambia, the native countries of some *Sorghum* and *Gossypium* species, drought-, disease- and pest-resistant landrace populations of small grain and industrial, leguminous, and vegetable plants were found. Drought-resistant wheat, barley and sorghum were obtained from Burundi and Tanzania, as well as samples of early ripening maize, *Pennisetum, Phaseolus* beans and other crops (Kobylyanskaya 1989).

The South American continent has always been of interest to the Institute's scientists, especially in recent times, from the point of view of collecting new, unconventional crops for the Soviet Union. Of late, the Institute's collecting missions have produced collections of *Stevia*, quinoa, jojoba and other crops. In the 1980s explorations were carried out in Bolivia (1980, 1988), Peru (1980, 1989), Brazil (1984, 1987), Colombia (1980), Argentina (1982), Venezuela (1987) and Ecuador (1989). During this period VIR carried out scientific visits to the USA (1986, 1989), Canada (1987) and Australia (1988). These visits provided a way of acquainting scientists with breeding activities and of collecting plant cultivars from those countries.

Thus in each year of the 1980s, the Institute organized collecting missions to between four and eight foreign countries, usually lasting 20 to 30 days on average. Participants in a collecting mission made thorough preparations for their collection work; they studied the geography, climate, vegetation, as well as scientific information on plant industry, plant breeding, genetics and botany. They visited scientific institutions and local markets of that country as well as collecting plants in the wild. For immediate seed exchange, they brought 50 to 150 of the best commercial varieties from the VIR collection with them.

When collections took place abroad, all crop species were usually collected rather than a specialized collection of a single crop. Collecting missions undertaken abroad added 4000-5000 samples annually to the VIR genebank. For example, in 1989, the USSR sponsored collecting expeditions to Poland (400 collected accessions of cereals, legumes, forages, etc.), China (310 accessions of cereals, legumes, vegetables, etc.), Bhutan (500 accessions of cereals, legumes, vegetables, etc.), Côte d'Ivoire (480 accessions of tropical species), the USA (a joint Soviet-American mission collected 315 accessions of wild relatives of sunflower), Bolivia (900 accessions of cereals, legumes, potato, cotton, etc.), Ecuador (800 accessions of potato, cereals, legumes, vegetables, amaranthus, etc.), and Peru (400 accessions of potato, cotton, maize, legumes, *Lupinus*, etc.).

1990s

In the 1990s the collecting activities of the Institute became less intensive owing to a reduction in VASKhNIL's funding of international programmes (Colombia – 1990; Tunisia, Egypt, Portugal – 1991; Costa Rica – 1992 and last seed mission to Canada – 1994). At the same time, long-term trips within the programme of exchange of scientists became more and more frequent, adding to the collection numerous crop accessions from the breeding centres of the United States (1991) and other countries.

In the course of VIR's activities after the Second World War, collecting missions and scientists travelling abroad have brought to the Institute from foreign countries approximately 200 000 plant germplasm samples. Specifically 14 129 accessions were introduced from 1945 to 1959, 17 442 from 1960 to 1969, 43 664 from 1970 to 1979, and 50 784 from 1980 to 1991 (Mesherov *et al.* 1982; Kobylyanskaya *et al.* 1990; Zoteeva *et al.* 1996).

With the purpose of botanical surveying, VIR's scientists have visited nearly all the world's countries where Vavilov and his scientific associates formerly carried out plant explorations, except for the following: Cyprus, China Taiwan, Lebanon, Cuba, Panama and Israel. All in all, 178 collecting missions have taken place abroad in more than 90 states during the entire post-war period (see Appendix IV). VIR's researchers have not only continued plant collecting in the countries previously visited by Vavilov and his followers, but have also travelled to and explored the territories of another 38 foreign states. This tremendous work has resulted in over 150 000 new accessions of seed samples and other plant germplasm, plus new scientific publications and personal contacts with foreign scientists (Alexanian 1997).

Following the pattern of the USA the division of the Department of Applied Botany and Plant Breeding established by Vavilov in America in the 1970s and 1980s, (see Chapter 3), VIR with VASKhNIL's support opened foreign reproduction stations in Mexico and Vietnam, thus establishing a new efficient means of adding endemic plant species of those countries to VIR's genebank. During the years covered by their activities, the Institute's collections acquired about 15,000 new samples of valuable plant species (Gorbatenko 1994).

Vavilov's theory of plant introduction combined with the subsequent developments introduced by his followers enabled scientists to classify the plant genepools stored at VIR into four groups:

1. Genetic diversity from centres of origin;
2. Landrace populations developed through folk breeding;
3. Modern breeding varieties;
4. Genetic lines, introduced mutants and other new forms experimentally obtained.

Many years of work have helped the scientists of the Vavilov Institute to collect and preserve the genepool of cultivated plants and their wild relatives collected in almost all the countries of the world's five continents. Vavilov assigned paramount importance to collecting plant genetic resources in their original habitat. This approach to replenishing the VIR genebank collections has remained the most important one to this day.

Organization of collecting and exchange

As we have seen, VIR's collection was enlarged through plant-collecting missions in the USSR and abroad, as well as by exchanging samples with both USSR scientific institutions and foreign ones. The Institute established the Plant Introduction Department for this purpose. The principal task of this Department was to organize the exploration and collection of cultivated plants and their wild relatives and the quarantine testing of material received. Every year the Plant Introduction Department developed preliminary lists of samples to be collected abroad through eight to ten collecting missions. The collecting missions programme was approved by the Scientific Board of the Vavilov Institute and then endorsed and sponsored by VASKhNIL. The Academy requested the country to be visited to admit the collecting expedition and, at the same time, forwarded to that country the scientific outline and draft route of the planned expedition.

VIR's collection was also enlarged through the exchange of samples. In 1989, 9000 samples from the Institute's collection were sent to 46 countries, and VIR received 9500 samples from 45 countries. For this purpose the Institute published the seed exchange catalogue Delectus Seminum every three years and received similar catalogues from the USSR botanical gardens and foreign institutions. The Institute also satisfied requests for plant samples in addition to those offered through the catalogue. New varieties of agricultural crops that had been commercialized in the previous two years were the only exceptions. They were not sent to foreign contacts but other seeds were dispatched free of charge for scientific purposes.

All the plant materials, obtained by collecting missions or from other sources, underwent testing at the Quarantine Laboratory which was supervised by the USSR State Quarantine Inspection. After laboratory quarantine and fumigation, all samples were registered at the Plant Introduction Department. Those collected in the USSR were submitted immediately after registration to the various departments of plant resources.

Seeds, scions and seedlings received from foreign countries were sent for field quarantine testing to the Introduction Quarantine Nurseries (IQN) of seven VIR experimental stations that were located in different climate zones. All annual crops were tested at the IQN for a year, perennial herbaceous plants for two years, and fruit tree samples for three years. After that, the original seed and their progeny were sent to the departments of plant resources for detailed study and the fruit tree samples were dispatched as scions or seedlings (Bakhareva 1990).

Nowadays the interest of the Institute's scientists in the centres of origin of cultivated plants has not diminished. These centres continue to be regarded as the foci of maximum concentration of species and cultivar diversity capable of supplying agricultural and breeding needs for valuable plant traits and properties. The current strategy is to go on with a targeted search for such plant materials, adjusted to reflect current priorities.

Vavilov maintained that the problem of new crops and enhanced use of wild plants both in this country and abroad was especially important. With this in mind, the Institute has obtained and studied valuable plants previously unknown to national agricultural science and practice. Special attention has been paid to jojoba (*Simmondsia californica* Link.), an oil-producing plant recently introduced into cultivation abroad, which grows in arid regions of Mexico and the USA. Quinoa (*Chenopodium quinoa*

Willd.), about which N.I. Vavilov wrote is also very interesting for Russia. This plant grows in the South American Andes and is widely used for bread and forage production. Also promising is amaranth, a plant of universal applications, with a higher ascorbic acid content than that of spinach.

From Western and Central Africa, the Institute obtained voandsea groundnut (*Voandsea subterranea* Thouars.), a potentially valuable leguminous shrub, which resembles groundnut in its habit. It is a drought resistant plant capable of growing under high temperatures on poor sandy soils where soyabean and groundnut cannot be cultivated. Its seeds contain 18% protein and 68% carbohydrates, and yield 0.9 to 1.0 tonnes per ha.

In Central Asia a new crop, blue millet (*Panicum antidotale* Retz.), notable for its drought resistance and green matter yield, is being introduced into cultivation. A number of other plants have been introduced into Russia and undergone study. These include typhon (*Brassica* sp.), a new forage crop obtained by hybridization of Chinese cabbage and turnip and demonstrating cold tolerance and rapid growth (producing up to 50 tonnes of green matter per hectare after one to three mowings), guayule (*Parthenium argentatum* A.Gray.), a natural rubber-producer; and velvet grass or stevia (*Stevia rebaudiana* Bert.) which produces sugar. Work has been launched to introduce petroleum-producing plants, the sap of which may be used as a natural substitute for mineral oil, halophyte plants tolerant of high soil salinity (for instance, oil-bearing *Salicornia* sp., spurge and watermelon), rare citruses and rare fruits, such as babako (*Carica pentagona* Heilb.) and borojoa (*Borojoa patinoi* Cuatr.). There is interest in growing alternative forage plants, such as perennial resin-weed (*Silphium perfoliatum* L.) yielding over 100 tonnes of green matter per hectare, *Galega orientalis* Lam. with high protein content, and others.

The VIR collections served as the base for introduction into Russian agricultural practice such plants as amaranth, Beijing cabbage, myrobalan plum, honeysuckle, sea buckthorn, and *Actinidia*. In the coastal regions of the Caspian Sea, new arid forage crops have been introduced, such as winterfat (*Eurotia ceratoides* L.), *Kochia scoparia* L. and saxaul (*Haloxylon persicum* Bunge) (Gorbatenko 1994).

The Institute has maintained wide international relations, cooperating with many countries in Europe, America, Africa, Asia, Australia and Oceania. This cooperation has involved the exchange of initial material for breeding, the exchange of scientific literature, joint research programmes, and the exchange of scientific delegations. VIR scientists have strongly supported professional contacts because they believed that there was a real danger of losing many old and local and even new cultivars, as well as many wild relatives of cultivated plants. Many of these would have valuable traits and yet undiscovered potential. They all agreed that these precious resources ought to be preserved in genebanks for the sake of future generations.

8 Creation of the network of genebanks in the COMECON countries[7]

Establishment of the Scientific Technical Council (STC) for collections of wild and cultivated plants

In 1962, the 5th Conference of COMECON countries, dedicated to coordination of scientific activities in agriculture and forestry, adopted a decision which made the Institute of Plant Industry (VIR), responsible for convening a Session of the Coordination Council of representatives of the COMECON countries. It was resolved that such a Session was necessary for discussing the problems of expanding collaboration in joint investigations and practical measures towards more efficient collecting, exchange, study and utilization of crop genetic resources.The First Session of the Coordinating Council devoted to that problem was held in April 1964, in Leningrad. Germplasm experts representing national Plant Genetic Resources programmes from the USSR, Bulgaria, Hungary, East Germany, Mongolia, Poland, Czechoslovakia and Yugoslavia, as well as specialists from Romania, participated in this Session.

The participants in the session outlined explicit working plans and agreed on a joint scientific project entitled "Collecting, Studying, Preservation in Viable Conditions and Utilization of Global Resources of Cultivated and Wild Plant Species for Breeding Crop Varieties and Hybrids with Higher Yield and Quality". VIR, having the richest plant genetic resources collection and well-developed, scientifically-based methods of collecting, conservation and study, was appointed as coordinator of this cooperation. The countries participating in the international cooperation nominated national co-executives-in-charge. The genetic resources network of the COMECON countries incorporated the following institutions: Institut po Introductia I Rastitelni Resursi, Sadovo (the Institute for Plant Introduction and Plant Resources, Bulgaria), Agrobotanikai Intézet, Tapiószele (Centre of Agrobotany, Hungary), Zentralinstitut für Genetik und Kulturpflanzenforschung, Gatersleben (Central Institute of Crop Genetics and Research, Germany), Negdliin Urgamal Gazar Tarialangiin Erdem Shinzhilgeenii Hyreelen, Dorhan, (Institute of Plant Production and Cultivation, Mongolia), Instytut Hodowli i Aklimatyzacji Roślin, Radzików (Plant Breeding and Acclimatization Institute, Poland), Institutul de Cercetări rentru Cereale şi Plante Tehnice, Fundulea (Institute of Cereal and Industrial Crops, Romania), and Výzkumný ùstav Rostlinné Výroby, Praha-Ruzyně (Research Institute of Plant Production, Czech Republic).

In 1964, the national collections of cultivated plants available in the collaborating institutes were rather insignificant. Even VIR's collection, the largest and richest of them, could not completely provide the breeders with the valuable scientific research in their countries. After this, COMECON member countries enhanced and systemized the exchange of plant genetic resources between them. The first decade of joint research brought positive results: national collections of the member countries were enlarged by nearly 200 000 accessions of various crops. The next new step in the cooperation of the COMECON countries was taken when the 34th Session of the COMECON Permanent Commission for collaboration in the sphere of agriculture, held in October 1972, adopted a

[7] The establishment of genebanks in the COMECON countries is described in greater detail in Appendix 7.

resolution on establishing the Scientific Technical Council for Collections of Wild and Cultivated Plants (STC). The STC with membership consisting of two representatives from each of the collaborative institutes was set up with the purpose of further expanding cooperation in this field (Alexanian 1989).

The basic objectives of the STC were discussed and adopted at the first STC meeting held in Leningrad in 1973. These objectives included:

- organization of plant genetic resources collecting and plant genetic resources exchange for expanding the national collections;
- planning of explorations in order to collect germplasm of cultivated plants and their wild relatives;
- development of unified descriptors for diverse agricultural crops;
- application of computers in plant genetic resources registration and documenting the results of plant genetic resources studies;
- studying collection accessions and supplying germplasm to the research and breeding institutions of partner countries;
- development of measures for long-term preservation of collections of concern;
- systematic exchange of scientific information between collaborating institutes (Dorofeev 1987).

The STC held a Session each year in one of the countries included in the cooperation project with the goal of summarizing the year's activities and making plans for the future. Working conferences of specialists were also held on a yearly basis with the purpose of solving individual problems, exchanging experiences and discussing jointly-developed descriptors.

The world crop collection of VIR was approved as the central genebank of the COMECON countries. In accordance with the working plans and recommendations of the STC, the institutes replenished their collections by exchanging seed samples and other germplasm and by collecting plant genetic resources within their countries and abroad. Only those accessions that could be used in national breeding programmes, were preserved and studied in the national genebanks along with domestic landraces and wild species.

The EUCARPIA Genebank Committee, on which representative of all the COMECON countries, including the USSR, took part held meetings in each of the genebanks year by year, including West European countries. Scientific discussions on genetic resources also took place at each of these meetings concerning all aspects of genebank collecting, management and utilization.

Organization and execution of joint collecting missions

Joint expeditions were orgnized throughout the territories of the USSR and Eastern European countries, and germplasm exchange was carried out systematically, based on the requests of the countries. Thus, joint explorations took place in Poland (1976, 1978, 1980), East Germany (1980) and the Soviet Union (Georgia in 1977 and Krasnodar Region, Stavropol Region, North Ossetia, Daghestan (Russia), Azerbaijan and Georgia, in 1981). These collecting missions were staffed with representatives of Poland, East Germany, Czechoslovakia and the Soviet Union. They resulted in delivery to the national collections of the participating countries of over 3,500 samples of mainly local varieties and populations of cereal, leguminous and forage crops and their wild relatives. During this period, representatives of Romania, Hungary and Mongolia did not take part in international collecting missions in other countries, but at the same time they activated collecting of local varieties in their own countries (Alexanian 1997).

In 1983, the ninth meeting of the STC convened in Hungary adopted a plan of collecting missions in member countries for 1984 to 1990. It was from that time that regular joint collecting missions were carried out in the COMECON states on bilateral and multilateral bases. However, for some institutes, the major source of added new germplasm to the national collections remained the exchange of seed samples and planting materials between COMECON member countries. The first regular collections of local cereal and forage crop cultivars were performed in 1984 (Poland) and 1985 (Mongolia) by specialists from East Germany together with representatives of the countries where these explorations took place. In the following years, multilateral collecting missions became more prevalent. In 1986, two explorations were held: in Bulgaria with the participation of Hungary and the USSR, and in the Soviet Union (Georgia, Armenia and Azerbaijan) with the participation of Poland and Czechoslovakia. In the following year, there were three such collecting missions: in Mongolia (together with East Germany), Poland (jointly with Czechoslovakia and Bulgaria) and Czechoslovakia (jointly with Poland, Bulgaria and Hungary).

In 1988, a complex joint mission collected germplasm of the *Allium* family in the USSR in Uzbekistan, Tadjikistan, Kirghizia and Kazakhstan, with the participation of Czechoslovakia, Bulgaria, Poland and the USSR. In the same year, two minor explorations took place in Bulgaria and East Germany with the participation of Czechoslovak representatives. In 1989, onion germplasm collecting was continued in Turkmenia, Uzbekistan and Kazakhstan, with the participation of Polish specialists, and two collecting missions specializing in *Aegilops* were accomplished in Bulgaria (with the participation of the USSR and Poland).

In 1990, collecting missions with the participation of Czechoslovakia, Bulgaria and the USSR surveyed Poland and Mongolia where they collected local leguminous and forage crops. In the same year, in the USSR, Czechoslovakia and Poland participated in the next phase of the *Allium* collecting project in Altai and West Siberia (Russia), while Czechoslovak, Polish and Bulgarian specialists joined the mission which collected fodder grasses and their wild relatives in Kazakhstan and Uzbekistan.

All in all, in the period from 1986 to 1990, 16 collecting missions delivered over 7000 accessions of cultivated crops and their wild relatives (see Appendix V). Thanks to the STC's activities and the efforts of all member countries in plant genetic resources collecting, from 1964 the germplasm collections increased 1.5 times in Czechoslovakia, twice in Hungary, three times in Bulgaria and the GDR, and seven times in Poland.

Joint field evaluation and use of plant genetic resources in plant breeding

This enormous plant genetic diversity constituted an immeasurable treasure ensuring the successful implementation of the most advanced plant breeding programmes of the COMECON countries. Within the COMECON framework, the member countries carried out a comprehensive ecological study of the germplasm used as a source for plant breeding. This work was coordinated by the STC and supervised by VIR. The germplasm underwent assessment in all the member countries and at VIR's Experimental Stations under various ecogeographical conditions. Such assessment resulted in identification of promising materials for breeding widely adapted to the environments of Eastern Europe or the European or Asian parts of the USSR. The results of collecting missions and evaluation of plant genetic resources were published under the title *The Evaluation of Plant Genetic Resources and their Use in Breeding* (1987, 1990) '

As well as the above research on plant materials for breeding, an International

Agreement on the problem of "Development of Theoretical Principles for Plant Breeding and Seed Production and New Methods for Breeding of High Yielding Crop Cultivars and Hybrids" was signed in 1971 by the COMECON members, and the Coordination Centre (COC) headed by the All Union Institute of Plant Breeding and Genetics (Odessa, Ukraine). In the framework of this programme, along with the plant varieties received from different countries, large sets of hybrid populations were studied at sites with contrasting climatic conditions. Basic crops for such studies were wheat, barley, maize and sunflower. Complex studies of the materials stored in the national collections of COMECON member countries enabled them to identify new and valuable materials for breeding which formed the core of many crop cultivars widely used not only in their countries of origin, but also in other cooperating states.

On the basis of the vast amount of initial plant germplasm recommended and submitted to the plant breeding institutions of the COMECON countries, about a hundred new cultivars and hybrids of various agricultural crops were created each year. From 1967 to 1989, about 800 samples and breeding lines of wheat, rye, barley and triticale were tested in ecological experiments supervised by COC (1983,1985,1987,1990). As a result of successful programmes, 750 new cultivars were released for state variety trials, and over 350 cultivars of cereals were commercialized. International collaboration resulted in cooperatively bred cultivars of winter wheat (3 cultivars), winter barley (2 cultivars) and spring barley (2 cultivars) and winter rye (4 cultivars).

In the sphere of genetic resources management, results of studies on preservation and other matters were discussed at the STC sessions, meetings of the COC working groups and other international conferences. In 1988, due to social and political changes in the socialist states of Eastern Europe, the STC decided at its regular meeting to develop new and more efficient forms of cooperation. At that time in its work, the STC paid a lot of attention to the development of a unified documentation system in order to facilitate exchange and accumulation of information on various issues related to plant genetic resources including methods of seed conservation and coop evaluation

On the basis of the descriptors worked out at VIR since 1973, international descriptor lists were developed for most of the agricultural crops. This undertaking was initiated in 1974 when the international descriptor list was issued for the *Triticum* genus. Later, these COMECON descriptor lists were used with othersd to establish the descriptors developed by IBPGR/IPGRI for various crops. Many years of collaboration resulted in 42 descriptor lists (see Appendix VI) which had been worked our for major agricultural crops. They were meant to be used for studying the germplasm accessions from COMECON countries and enabling the compilation of a comparative data system, on which decisions on germplasm exchange could be based, facilitating selection of promising materials for plant breeding.

With the purpose of development and introduction of mathematical data processing methods and information retrieval systems for national databases, an Acronym Descriptor List (list of abbreviations) was devised. It provided for the creation in 1990 of a unified database for wheat, encompassing 8500 accessions. Steps were also taken towards establishing an international database of the COMECON countries and a genetic resources information retrieval system.

Joint activities in establishing databases on crop collections made it possible to enhance and improve this kind of research in each member country. For example, Czechoslovak scientists were successful in developing and introducing the EVIGES information system on plant genetic resources which has been fruitfully applied up to the present. Similar information systems were worked out in East Germany, Poland and later in Hungary.

Joint research contributed to further improvement in methods and technologies of long-term plant genetic resources storage in national collections. Before COMECON was dissolved, the accessions in Bulgaria, Hungary, Poland, Czechoslovakia and the USSR could be stored at temperatures below zero for a long period of time. About 300 000 seed samples were already stored in the national genebanks by this time. As the leading institute, VIR worked out the structure of the International Centre for Plant Genetic Resources of the participating countries, which was duly adopted by COMECON headquarters. However, further political processes in the East European countries and winding up of the COMECON system finally ended this form of cooperation.

Considering the aggregate genepool preserved by the COMECON member countries as the heritage of all mankind, at the STC's closing session in 1990, representatives of the national genebanks sent a petition to FAO requesting that the programme of East European COMECON countries, established in the course of almost 30 years of collaboration, be regarded as a regional programme and thus able to receive all kinds of support, including financial aid, in order to prevent destruction or abrupt reduction of the valuable plant genetic diversity accumulated by joint efforts.

Such a programme of cooperation resulted in the establishment and development of East European national programmes and, in the 1980s, genebanks for long-term storage of plant germplasm were organized in Bulgaria, Poland, Czechoslovakia and East Germany These were based on Vavilov's concept concerning the accumulation and conservation of plant genetic resources, belonging mostly to local genepools, for their further study and utilization in national breeding programmes (Alexanian 1997). This complex programme on plant genetic resources collecting, preservation and study, which united the efforts of numerous research institutions and individual experts of the former COMECON countries provided a great contribution to the theory and practice of agricultural production in Eastern Europe.

9 Conclusions – VIR's activities after 1990

The dramatic changes that have taken place in the former Soviet Union since 1990 have clearly influenced the Institute and its programmes. While the responsibilities of VIR have remained the same, the methods and approaches required to fulfil them must now be completely altered. In 1992 six experimental stations which had been within the VIR's network and had maintained regional collections of various crops along with rich collections of fruit, subtropical crops and vineyards, became within countries of the CIS (Commonwealth of Independent States), namely Ukraine, Uzbekistan, Kazakhstan, Georgia and Turkmenia. Scientific contacts with these stations are maintained, but there is no proper mechanism for the exchange of plant genetic resources, though exchange is going on to some extent. Intergovernmental agreements regarding plant genetic resources are being devised which, it is hoped, may result in linking together the scientific programmes of the CIS countries.

The loss of experimental stations impedes the further collecting of plant diversity by VIR's expeditions in the territories of the above countries due to the political and economic situation in the countries, the rise of nationalism and outbreaks of national and ethnic conflict, as well as the absence of specific laws concerning natural genetic diversity. Political and military disturbances in several countries of Central Asia and Transcaucasus have prohibited regional expeditions. Organization of plant collecting within Russia and the CIS countries is considerably limited by the lack of funds and high inflation rates in all countries of the former USSR.

The allocation of funds for sending requests for seeds to foreign countries and mailing seed samples to meet the incoming requests has almost stopped. There is no financing for purchases of machinery and equipment for the Institute and experimental stations. Funds for research work are scanty and salaries stay low while the available funds are constantly diminished by inflation. To the merit of the majority of researchers and technical assistants at the Institute and experimental stations, they are committed to their task and make the maximum effort to preserve and study the global collections held at VIR.

In 1994, the total collections of the Institute numbered 334 000 accessions belonging to 86 botanical families, 2102 species of 425 genera. Half of this diversity is preserved in the National Seed Store (Kuban, Krasnodar Territory) the renovation of which has been sponsored by USDA and IPGRI since 1994. This Project involves replacement of the refrigeration plant and development of a computerized database for VIR's collections. Joint efforts are envisaged for constructing a modern genebank in St. Petersburg providing long-term storage of accessions in addition to those in Krasdnodar.

Modern facilities are to be built in St. Petersburg for storing the base collection at –20°C because of the expense involved with the annual regeneration of about 100 000 samples needed to preserve seed germination at the experimental stations. Activities on organizing *in vitro* culture storage have gained a new impetus, and experiments with cryopreservation of plant materials are being carried out.

The number of publications summarizing results of studies of the collections has become reduced. At the same time, preparation of the second edition of the *Theoretical bases of breeding* was started in 1990. In 1993 the volume entitled *Molecular biology aspects of applied botany, genetics and breeding* was published, in 1994 *Physiological aspects of applied botany, genetics and breeding*, and in 1995 *The genebank and breeding of grain legumes*. A new volume of *The*

Cultivated Flora devoted to soyabean is underway.

At present, in cooperation with other research institutes of the Russian Academy of Science (RAN) and the Russian Academy of Agricultural Science (RASKhN), a new scientific project seeks to identify genes that are available worldwide, determine new priorities in crop research, study the possibility of combining valuable genes in one genotype, and create donors for breeding programmes. VIR must find new areas in which to arrange alternative reproduction and testing sites with climate and ecological conditions similar to the conditions of those research stations no longer available as a result of the breakup of the former Soviet Union.

It is important to stress that the Institute's staff are well aware of its historical and practical responsibilities regarding the safety of the irreplaceable collections held by VIR. The new generation of Vavilov's successors will neither abandon VIR nor leave the Institute's plant collections vulnerable to damage and destruction. Despite our current problems, Vavilov's spirit continues to instill confidence in our own abilities and knowledge.

In spite of difficulties, the Russian Government was able to provide VIR with 90 ha of land near Adler City (Krasnodar region) to establish an experimental station of subtropical and southern crops within Russia. Nine hundred hectares of land have been transferred to VIR in the dry subtropical zone of Daghestan at the border with Azerbaijan. Over 1000 ha have been allocated for the collection of plant genetic resources by the municipal authorities of Volgograd (Volga river basin). In Gelendzhik City (Black sea), a new experimental area is now being established for the grape collection. VIR has also started organizing four new quarantine nurseries in addition to those belonging to the six stations that remain in VIR's network. VIR regards international cooperation in science and technology as the most important aspect of its activities and abides by its international obligations.

The activities of VIR within the ECP/GR framework should be noted, for this programme is paying special attention to the preservation and study of European genetic resources. Experts from VIR participate in six working groups (*Avena, Hordeum, Beta, Allium, Brassica,* Forages and Sunflower). Within the framework of the programme, a joint Russian-German collecting mission targeting *Beta* species was carried out in 1994. On the basis of an agreement signed by VIR and the US Department of Agriculture/Agricultural Research Service, a bilateral governing committee has been organized with a goal of coordinating and targeting their joint activities. Of particular significance is an agreement between VIR and IPGRI which has been operative since 1990.

An important event in the Institute's history took place in August 1994. When the Vavilov Institute of Plant Industry celebrated its centennial with a five-day international symposium. "Global Plant genetic Resources – Heritage of Mankind". It concluded with a visit to the VIR laboratories, the Pushkin Experiment Fields, The Pavlovsk Experimental Station, and other agricultural research facilities in the area. In numerous publications dedicated to this event and in the addresses at the Conference it was mentioned that all activities of the Institute abide by Vavilov's principles in applying an integrated approach to collecting, studying and preserving plant genetic resources. It was reported that the revival of Vavilov's name and ideas in the post-war period helped to launch extensive and detailed explorations of global plant diversity. New data received by collecting missions were made the basis of further development of the theory of centres of crop origin undertaken by the Institute's researchers. The data derived from a comprehensive approach to the study of the vast specific diversity of cultivated plants and their wild relatives made it possible to develop new or revise existing taxonomic systems covering major agricultural crops. These efforts resulted in the publication of new volumes of the Cultivated Flora started by Vavilov. Profound research on intraspecific variability helped to find or create plant forms predicted by the Law of Homologous Series in Variation.

Characterization and evaluation of the whole diversity preserved in the germplasm collections, which originated with a series of geographical expeditions, is now going on in various ecogeographical regions of the country, with the help of VIR's Methodological Guidelines, COMECON International Descriptor Lists and recently developed VIR Descriptors.

Those plans which Vavilov failed to accomplish in full became the priorities of the Institute's activities. These principles were used in working out the programme of identification and creation of sources and donors of important commercial traits. They were also used to establish character-orientated and genetic stock collections with identified genes, which may serve as a keystone in developing core collections of various crops. A complex approach was also applied for the preservation of the Institute's collections. The work with the collections is based upon a thoroughly developed botanical classification of each genus down to the level of botanical varieties. Much attention has been paid to the passport data of each new accession, so that duplicates could be detected as soon as possible.

The geographical principle of germplasm storage at the Institute's stations has remained dominant since Vavilov's times. According to an ecogeographical classification, all collection accessions are distributed to VIR's stations and placed in duplicate collections for reproduction and preservation. Two parallel ways of germplasm storage – long-term in the National Seed Store based on the principles worked out before World War II, and short-term in working and base collections at the experimental stations and the Institute itself, provide for the possibility of complementary *in situ* and *ex situ* conservation of germplasm accessions.

In the modern post-Soviet period, the Vavilov Institute regards closer cooperation with the world community as the best way of solving the problems facing the Institute. VIR has been – and still remains – the heritage of humankind. The importance of VIR's tasks in preserving global collections of plant genetic resources, many of which are now unavailable in nature and agricultural production, as well as its scientifically grounded traditions and the presence of highly qualified personnel may hopefully determine ways for the Institute to overcome critical times in its work in the near future.

The Institute's collection, which originated as long ago as the end of the 19th century, is now acquiring strategic importance. Considered from the beginning as a collection of Russian crop cultivars, it has grown to be one of the most comprehensively studied, systematic and representative collections of plant genetic resources in the whole world. The vital significance of this collection for agricultural production in Russia and the world cannot be doubted, and its importance will increase on a global scale in the light of the Convention on Biological Diversity, signed in 1992 in Rio de Janeiro by national and governmental leaders of the majority of the world's nations.

Appendix I. Major species collected by Vavilov on his collecting missions, 1916-1940

[All species are represented by numerous botanical varieties]

Abutilon avicennae Gaertn. – Chinese jute
Achras sapota L. – sapodilla
Aegilops spp. – wild wheat grass
  Aegilops crassa Boiss. – wild wheat grass
  Aegilops cylindrica Host. – wild wheat grass
  Aegilops squarrosa L. – wild wheat grass
  Aegilops triuncialis L. – wild wheat grass
Agave spp. – agave
  Agave atrovirens Karw. – maguey agave, (= *A. americana)*
Agropyrum tenerum Vasey. – wheat grass
Agrostis alba L. –
Aleuritres fordii Hemsley – tung oil tree
Allium spp.
  Allium cepa L. – onion
  Allium chinense Don. – perennial Chinese onio, tsyu-tsai
  Allium fistulosum L. – Japanese leek
  Allium macrostemon Bge. – syao-suan
  Allium porrum L. – leek
  Allium sativum L. – garlic
  Allium xiphopetalum Aitch et Baker. – wild onion
Amaranthus paniculatus L. – purple amaranth
Ammi copticum L. – azhgon
Amygdalus communis L. – almond
  Amygdalus fenzliana Lipsky – wild
  Amygdalus georgica Desf. – almond
  Amygdalus orientalis Mill. – wild
Andropogon contortus L. – weed
  Andropogon ischaemum L. – weed
  Andropogon halepensis Brot. – weed
  Andropogon laniger Desf. – wild
  Andropogon sorghum Brot. – sorghum
Anethum graveolens L. – dill
Annona cherimola
Apocynum venetum L. – dogbane
  Apocynum cannabium L. – dogbane
  Apocynum hendersonii Hook. – dogbane
Arachis hypogaea L. – peanut
Aralia cordata Thunb. – udo
Arctium lappa L. – burdock, Japanese godo
Areca catechu L. – arecanut
Armeniaca vulgaris L. – apricot
Arrhenatrum elatius M. et K.
Asclepias spp.
  Asclepias syriaca L. – milkweed, silkweed
Asparagus lucidus Lindl. – clubshaped tubercle asparagus
Avena abyssinica Hochst. – Abyssinian oat
  Avena barbata Pott. – wild oat
  Avena byzantina C.Koch. – red (Mediterranean) oat
  Avena fatua L. – weed oat
  Avena hirtula Lag. – wild oat
  Avena ludoviciana Dur. – weed oat
  Avena sativa L. – oat
  Avena sterilis L. – wild oat
  Avena strigosa Schreb. – sand oat
  Avena vaviloviana Mordv. – wild oat
  Avena wiestii Steud. – wild oat
Bambusa spp. – bamboo
  Bambusa miltis Poir. – bamboo
Berberis heteropoda Schrenk. – barberry
  Berberis integerrima Bge. – barberry
  Berberis orientalis C.K.Schneid. – barberry
  Berberis vulgaris L. – European barberry
Beta cicla L. – white beet
  Beta maritima L. – wild beet
  Beta vulgaris L. – beetroot, sugar beet
Bixa orellana L. – achiote or anatto
Boehmeria nivea Hook. et Arn. – ramie
Bomarea acutifolia Herb. – bomarea
Bouvardia ternifolia Schl. – bouvardia
Brassica rapa L. spp. *oleifera* Metzg. – fodder kale
  Brassica rapa L. spp. *rapifera* Metzg. – turnip
  Brassica carinata A.Braun. – leaf mustard

Brassica chinensis L. – Chinese cabbage
Brassica juncea Czern. et Coss- Indian or Chinese mustard
Brassica napiformis Bailey. – Napa cabbage
Brassica oleracea L. – cabbage
Brassica pekinensis Rupr. – chinese cabbage
Bromus inermis Leyss.
Cajanus cajan – pigeonpea
Calocarpus spp.
Calotropis procera – *C.gigantea* ait.
Cannabis indica Serebr. – wild
Cannabis ruderalis Janisch. – wild
Cannabis sativa L. – hemp
Capsicum annuum L. – pepper
Capsicum frutescens L. – perennial pepper
Carica papaya L. – papaya
Carthamus tinctorius L. – safflower
Carum spp.
Carum carvi L. – caraway or cumin
Carum sogdianum Lipsky. – cumin
Casimiroa edulis La Llave. – white sapote, Mexican apple
Cassia fistula L. – wild cassia
Castanea sativa Mill. – chestnut
Castanea vesca Gaeztn.- chestnut
Castilla elastica Cerv. – Panama rubber
Cerasus avium Mnch. – sweet cherry = *Prunus avium*
Cerasus vulgaris Mill. – sour cherry = *Prunus cerasus*
Ceratonia siliqua L. – carob
Chaenomeles lagenaria Koidz. – Chinese quince
Chenopodium spp.
Chenopodium ambrosioides L. – Mexican tea
Chenopodium nuttaliae Saff. – apazote
Chenopodium quinoa Willd. – quinoa
Chrysanthemum coronarium L. – chrysanthemum
Chrysanthemum morifolium Ram. – mum, edible chrysanthemum "shiso"
Chrysothamnus nauseous Britt.
Cicer arietinum L. – chickpea
Cichorium intybus L. – chicory
Cinchona spp.
Cinchona colisaya Wedd. – quinine tree
Cinchona cordifolia Mut. – quinine tree
Cinchona succirubra Pav. – quinine tree
Cinnamomum camphora L. – camphor tree
Citrullus colocynthis L. – wild water melon
Citrullus lanatus Mansf. – water melon
Citrullus vulgaris Schrad. – water melon
Citrus aurantium L. – bitter orange
Citrus sinensis Osb. – sweet orange
Citrus limonum Risso. – lemon
Coffea arabica L. – coffee
Colocasia antiquorum Schott. – taro
Coriandrum sativum L. – coriander
Cornus mas L. – cornel
Corylus avellana L. – hazelnut
Corylus colchica Alb. – Turkish hazelnut
Corylus colurna L. – Turkish hazelnut, Coconut
Cosmos bipinnatus Cav. – cosmos
Cosmos caudatus H.B.K. – cosmos
Cosmos diversifolius Otto. – cosmos
Cosmos sulfureus Cav. – cosmos
Cotoneaster aitchisonii C.K.Schneid.
Cotoneaster fontanesii Spach.
Crataegus spp.
Crataegus mexicana Moq. et Sesse. – Mexican hawthorne
Crataegus stipulosa Steud. – texocote
Croton tiglium L. – wild
Cucumis agrestis Pang. – wild
Cucumis chinensis Pang. – Chinese cucumber
Cucumis melo L. – melon
Cucumis melo L. var.flexuosus Naud. – snake melon
Cucumis microcarpus Pang. – weed
Cucumis prophetarum L. – globe cucumber
Cucumis sativus L. – cucumber
Cucurbita ficifolia Bouche. – fig-leaved gourd
Cucurbita maxima Duch. – giant squash
Cucurbita mixta Pang. – squash
Cucurbita moschata Duch. – cushaw
Cucurbita pepo L. – pumpkin squash
Cucurbita turbaniformis Alef. – turban squash
Cuminum cyminum L. – cumin
Curcuma zedoaria Rosc. – zedoary
Cydonia oblonga Mill. – quince

Cyperus esculentus L. – zulu nut
Cyperus papyrus L.- paper reed
Dahlia spp. – dahlias
Dahlia excelsa Benth. – dahlia
Dahlia imperialis Roezb. – dahlia
Dahlia variabilis Desf. – dahlia
Datura stramonium L. – thorn-apple
Daucus carota L. – carrot
Delphinium zalil Aitch. et Hemsl.
Dioscorea batatas Decne. – chinese yam
Diospyros spp.
Diospyros kaki L. – persimmon, kaki, chinese date plum
Diospyros lotus L. – date plum
Elaeagnus angustifolia L. – oleaster, wild olive
Elaeagnus hortensis
Elaeagnus orientalis L. – Russian olive
Eleocharis tuberosa Schult. – water chestnut
Eleusine coracana Gaertn. – finger millet
Ensete ventricosum (Welw.) Cheesman (syn. Musa ensete J.F.Gmel.) *Abyssinian* banana
Eragrostis abyssinica L. – teff = E. tef
Eriobotrya japonica Lindl. – loquat
Eruca sativa Lam. – garden rocket
Erythroxylon coca Lam. – coca
Euchlaena mexicana Schrad. – teosinte
Fagopyrum esculentum Moench. – buckwheat
Fagopyrun tataricum Gaertn. – Tatar buckwheat
Feijoa sellowiana Berg. – feijoa
Ficus carica L. – fig
Foeniculum officinale All. – wild fennel
Foeniculum vulgare Mill. – fennel
Glycine spp.
Glycine max Merr. – soy bean
Gossypium arboreum L. – cotton tree
Gossypium brasiliense Macf. – cotton
Gossypium herbaceum L. – cotton
Gossypium hirsutum L. – cotton
Gossypium mexicanum Todaro. – cotton
Gossypium peruvianum Cav. – cotton
Gossypium punctatum Schum. – wild cotton
Gossypium purpurascens Poir. – wild cotton
Gossypium vavilovii Proch. – wild cotton
Gossypium vitifolium Lam. – wild cotton
Guizotia abyssinica Cass. – noog, ramtil, niger seed
Hagenia abyssinica Willd. – kosso
Hedysarum coronarium L. – sulla
Helianthus annuus L. – sunflower
Helianthus lenticularis Dougl. – wild
Helianthus tuberosus L. – Jerusalem artichoke
Hevea brasiliensis Muell. Arg. – rubber tree
Hibiscus cannabinus L. – Indian hemp, bastard jute
Hibiscus esculentus L. – okra, gumbo
Hicoria pecan Britton. – pecan nut
Hippophae rhamnoides L.
Hordeum spp.
Hordeum bulbosum L. – wild barley
Hordeum caducum Mungo. – wild barley
Hordeum crinitum Coss. – wild barley
Hordeum humile Vav. et Bacht. – wild barley
Hordeum murinum L. – wild barley
Hordeum secalinum Schreb. – wild barley
Hordeum spontaneum C.Koch. – wild barley
Hordeum vulgare L. – barley
Ilex paraguayensis A. St. Hil. – mate tea
Ipomoea aquatica Forssk. – Chinese yung-tsai
Ipomoea batatas Lam. – sweet potato
Ipomoea heterophylla Ort. – morning glory
Ipomoea purga Wender. – morning glory
Ipomoea purpurea Lam. – morning glory
Ipomoea schiedeana Ham. – morning glory
Ipomoea tyriantha Lindl. – morning glory
Juglans fallax Dode – walnut
Juglans kamaonica Dode – walnut
Juglans regia L. – walnut
Lactuca sativa L. – lettuce
Lactuca oleracea L.
Lagenaria siceraria (Mol.) st. – bottle gourd
Lantana camara L. – lantana
Lathyrus cicera L. – lesser chickpea
Lathyrus sativus L. – chickling vetch or

grasspea
Lathyrus tinginanus L. – Tangier pea
Lens esculenta Moench. – lentil
Lens orientale Boiss. – Persian lentil
Lepidium sativum L. – garden cress
Lespedeza striata Hook. et Arn. – Japanese clover
Lilium spp.
Lilium tigrinum Ker. – lily
Linum indehiscens Vav. et Ell. – flax
Linum usitatissimum L. – flax
Lolium multiflorum Lam. –
Lolium perenne L. – perennial rye grass
Lucuma salicifolia H.B.K. – yellow sapote
Luffa acutangula Roxb. – luffa
Luffa cylindrica M.Roem. – edible luffa, sponge gourd
Lupinus mutabilis Sweet – Tarwi or Incalupin
Lycopersicon cerasiforme Dum. – cherry tomato
Lycopersicon esculentum Mill. – tomato
Lycopersicon peruvianum Mill. – Peruvian tomato
Malus spp.
Malus baccata Borkh. – Siberian crabapple
Malus domestica Borkh. – apple
Malus orientalis Uglitz. – wild apple
Malus pumila Mill. – paradise apple
Mangifera indica L. – mango
Manihot esculenta Crantz. – manioc / cassava
Medicago spp.
Medicago asiatica Sinsk.
Medicago falcata L.
Medicago lupulina L. – black medick
Medicago hemicycla Grossh.
Medicago polia Brand. – alfalfa
Medicago sativa L. – alfalfa, lucerne
Medicago syriaco-palaestinica Sinsk. – alfalfa
Medicago tetrahemicycla Sinsk. – wild medic
Melilotus spp.
Melilotus alba Medic. – sweet clover
Melilotus officinalis Lam. – sweet clover
Mentha spp. – mint
Mespilus germanica L. – medlar
Mirabilis jalapa L. – mirabilis
Momordica charantia L. – bitter gourd
Morus alba L. – white mulberry
Morus nigra L. – black mulberry
Musa ensete J.F. Gmel. – Abyssinian banana
Musa cavendishii Lamb. – banana
Musa sapientum L. – banana
Myristica fragrans Houtt. – wild nutmeg, mace
Myrtus communis L. – wild myrtle
Nardostachys jatamansi DC. – spikenard
Nelumbo nucifera Gaerth. – lotus
Nicotiana tabacum L. – tobacco
Nicotiana rustica L. – Aztec tobacco
Nigella sativa L. – black cumin
Ocimum basilicum L. – basil
Olea europaea L. – olive tree
Oryza sativa L. – rice
Oxalis tuberosa Molina – oca
Pachyrrhizus angulatus Rich. – jicama/yam bean
Papaver somniferum L. – opium poppy
Panax ginseng Mey. – ginseng
Panicum spp.
Panicum crus-galli L. – cockspur
Panicum frumentaceum Fr. et Sav. – Japanese millet
Panicum italicum L. var. moharicum – mohar
Panicum miliaceum L. – millet
Parthenium spp.
Parthenium argentatum A. Gray. – guayule
Parthenium incanum H.B.K. – mariole
Passiflora ligularis Juss. – sweet granadilla
Peganum harmala L. – wild "Turkey red"
Pennisetum typhoideum Rich. – bulrush, pearl millet
Perilla ocymoides L. – miso
Persea americana – avocado
Persica vulgaris Mill. – peach (= *Prunus persica*)
Petasites japonicus Mig. – sweet coltsfoot
Petroselium sativum L. – parsley
Phaseolus aconitifolius Jacq. – moth bean
Phaseolus acutifolius A. Gray – tepary bean
Phaseolus angularis Willd. – adzuki bean
Phaseolus aureus Pipper. – green gram
Phaseolus lunatus L. – lima bean

Phaseolus multiflorus Willd. – scarlet runner bean (= *P. coccineus*)
Phaseolus mungo L. – black gram
Phaseolus vulgaris L. – kidney bean
Philadelphus mexicanus Schlecht. – mock orange (= *P. coronarius*)
Phleum pratense L. – Timothy grass
Phoenix abyssinica – wild palm
Phoenix dactylifera L. – datepalm
Phyllostachys simonsoni Kras. – bamboo
Physalis ixocarpa – tomatillo
Pimpinella anisum L. – anise
Pimpinella griffithiana Boiss. – wild pistachio
Pistacia khinjuk – wild
Pistacia lentisens L. – pistachio
Pistacia vera L. – pistachio nut
Pisum arvense Moris. -
Pisum elatium M. B. – wild pea
Pisum fulvum Siebth. et Sm. – wild pea
Pisum humile Boiss. – wild pea
Pisum sativum L. – pea
Plantago ispaghula Roxb. – wild
Poa spp.
Poa pratensis L. – meadow grass
Polakowskia tacaco Pittier. – tacacco
Polianthes tuberosa L. – tuberose
Portulaca oleracea L. – purslane
Prunus armeniaca L. – apricot
Prunus capuli Cav. – black cherry
Prunus divaricata Lebed. – cherry plum
Prunus domestica L. – plum
Prunus padus L. – bird cherry
Prunus persica Batsch. – peach
Prunus salicina Lindl. – Japanese plum
Prunus simonii Carr. – apricot plum, Chinese plum
Prunus spinosa L. – blackthorn, sloe
Prunus tomentosa Thunb. – Chinese cherry
Psidium guajava L. – guava
Punica granatum L. – pomegranate
Pyrus spp.
Pyrus caucasica Fed. – pear
Pyrus communis L. – pear
Pyrus elaeagnifolia Pall. – snow pear
Pyrus salicifolia Pall. – snow pear
Pyrus serotina Rehd. – Chinese pear
Pyrus sinensis Lindl. – Chinese pear
Pyrus syriaca Boiss. – willow-leaved pear
Pyrus ussuriensis Maxim. – Ussurian pear
Raphanus sativus aestivus Alefeld. – radish
Raphanus sativus hybernus Alefeld.
Raphanus sativus raphanistroides Sinsk. – Japanese radish
Rheum palmatum L. – Chinese rhubarb
Ribes spp.
Ricinus communis L. – castor bean
Ricinus presicus Popova – castor bean
Ricinus sanguineus Horbtlorp.
Rosa damascena Mill. – Damascene rose
Rosa lutea Mill. – Austrian briar
Rosa moschata Mill. – moss rose
Rubia tinctorum L. – madder
Rubus spp.
Rubus anatolicus Focke.
Rubus caesius L.
Saccharum officinarum L. – sugarcane
Sagittaria sagittifolia L. var.sinensis Makino. – arrowhead (= *S. chinensis* Sims.)
Salvia chia Fernald. – chia
Salvia hypoleuca Benth. – wild
Secale cereale L. – rye
Secale montanum Guss. – wild rye
Secale vavilovii Grossh. – wild rye
Sechium edule Swartz – chayote
Sesamum indicum L. – sesame
Setaria italica L. – foxtail millet
Setaria italica L. subsp.maxima Al.
Setaria italica L. subsp.mocharica Al.
Sicana odorifera Naud. – coroa
Sinapis arvensis L. – rapeseed/wild mustard
Sisymbrium Sophia L.
Solanum spp. – potato
Solanum ajuscoense Buk. – wild potato
Solanum andigenum Juk. et Buk. – Andean potato
Solanum antipoviczii Buk. – wild potato
Solanum boyacense Juk. et Buk. – cultivated potato
Solanum caniarense Juz. et Buk. – cultivated potato
Solanum candelarianum Buk. – wild potato
Solanum coyoacanum Buk. – wild potato

Solanum cuencanum Juz. et Buk. – cultivated potato
Solanum demissum Lindl. – wild potato
Solanum goniocalyx Juz. et Buk. – cultivated potato
Solanum kesselbrenneri Juz. et Buk. – cultivated potato
Solanum melongena L. – eggplant
Solanum tuberosum L. – potato
Solanum vavilovii Juz. et Buk. – wild potato
Solidago laevenworthii Torr. et Gray – goldenrod
Sorghum cernuum Host. (Sorghum vulgare Pers.) – sorghum (= S. bicolor)
Sorghum durra Stapf. – (= S. bicolor)
Spinacia oleracea L. – spinach
Spinacia tetranda Stev. – weed
Spondias mombin L. – mombin
Spondias purpurea L. – hog plum, spanish plum
Stachys sieboldii Miq. – Chinese artichoke
Stizolobium hassjo Piper et Tracy – Yokohama bean
Tagetes spp. – marigold
Tagetes erecta L. – marigold
Tagetes lucida Cav. – marigold
Tagetes patula L. – marigold
Tagetes signata Bartl. – marigold
Thea sinensis L. (syn. Camellia sinensis L.) – tea
Theobroma cacao L. – cocoa tree
Tigridia pavonia Ker-Gawl. – cacomite, tiger-flower
Trapa bicornis Osb. – horn nut
Trapa bispinosa Roxb. – Singhara nut
Trifolium alexandrinum L. – berseem
Trifolium hybridum L. –
Trifolium pratense L. – clover
Trifolium repens L. – white clover
Trifolium resupinatum L. – persian clover
Trigonella foenum-graecum L. – fenugreek
Triticum aestivum L. – Wheat
Triticum aethiopicum Jakubz. – Abyssinian wheat
Triticum compactum L. – club wheat
Triticum dicoccum Schuebl. – emmer
Triticum dicoccoides Koern. – wild emmer
Triticum durum Desf. – durum wheat
Triticum persicum Vav. ex Zhuk. – Persian wheat
Triticum polonicum L. – Polish wheat
Triticum macha Dekapr. et Menabde. – masha wheat
Triticum monococcum L. – einkorn
Triticum spelta L. – spelt
Triticum timopheevii Zhuk. – Timopheevi wheat
Triticum turgidum L. – rivet or cone wheat
Ulex europaeus L. – gorse
Ullucus tuberosus Lozaro – ulluco
Vaccinium myrtillus L. – bilberry
Vaccinium uliginosum L. – bogwhortleberry
Vaccinium vitis-idaea L. – cowberry
Vanilla fragrans Ames. – vanilla
Vicia articulata Hornem. – single-flowered vetch
Vicia ervilia Willd. – bitter vetch
Vicia faba L. – faba bean
Vicia narbonensis L. – narbonne vetch
Vicia pliniana (Trabut) Murat. – wild
Vicia sativa L. – fare, vetch white
Vicia villosa Roth. – large Russian vetch, hairy vetch
Vigna catjang Walp. – cowpea, cajang
Vigna sinensis End. – horsegram
Vitis amurensis Rupr. – amur grapevine
Vitis vinifera L. – grape
Wasabia japonica Matsum. – Japanese horse-radish
Yucca aloifolia L. – yucca
Yucca elephantipes Regel. – yucca
Zea mays L. – maize, corn
Zinnia spp. – zinnia
Zinnia elegans Jacq. – zinnia
Zinnia mexicana Hart. – zinnia*Zinnia multiflora* L. – zinnia
Zizania latifolia Turcz. – zizania, wild rice
Zizyphus vulgaris Lam. – jujuba

Appendix II. People mentioned in the text

[This list, despite being incomplete in some details, is provided for the interest of the reader.]

Aaronsohn, A. (1876-1919) American botanist, who in 1906 discovered wild wheat in Syria and Palestine; Director of an experimental station in Jaffa.
Akemine, M. (?) Japanese botanist and plant breeder, professor in the University of Sapporo.
Alberts, H.W. (?) American plant breeder, expert in northern agriculture, Director of Alaska Station.
Alefeld, F. (1820-1872) German botanist, expert in cultivated and medical plants of Europe.
Alefeld, F.C. (1820-1872) Swedish botanist, plant taxonomist, expert in cultivated and medical plant of Europe.
Antropova, V.F. (1891-1972) Russian agricultural botanist, crop scientist, expert in rye; from 1925 worked at the Institute of Applied Botany.
Ascherson, P.F.A. (1834-1913) German botanist, florist, explorer of Mediterranean flora.
Atterberg, A.M. (1846-1916) Swedish botanist, taxonomist; was professor in chemistry at Uppsala University, later he was in charge of Chemical Station in Kalmar (Sweden). He published a number of articles concerning the identification of Nordic local varieties and about the adventive flora around Kalmar.
Audricour (?) French scientist, who worked in the Soviet Union.
Bakhareva, S.N. (?) Russian agricultural botanist, crop scientist, expert in tropical plants; was Head of the Department of Introduction VIR.
Bakhteyev, F.Kh. (1905-1982) Russian botanist, crop scientists, expert in barley; worked in VIR and in the Main Botanical Gardens of the USSR Academy of Sciences; Vavilov prize winner.
Barulina, E.I. (1896-1957) Russian agronomist, expert in legumes; from 1921 worked at the Institute of Applied Botany, was Vavilov's second wife.
Batalin, A.F. (1847-1896) Russian botanist, from 1870 worked at St.Petersburg botanical garden, from 1894 was the first Head of Bureau of Applied Botany.
Bateson, W. (1861-1929) British biologist, one of the fathers of genetics; from 1910 was the Director of the John Innes Horticultural Institute in Merton near London.
Baur, E. (1875-1933) German botanist and geneticist; in 1911-1927 was a professor of the Kaiser Wilhelm Biological Institute in Dahlem, Berlin; founder and director of the Institute of Plant Production and Genetics in Muenchenberg near Berlin.
Bazilevskaya, N.A. (?) Russian botanist, taxonomist, crop scientist, expert in oil and ornamental plants; from 1929 worked at the Institute of Applied Botany.
Beaven, R.E. (1857-1941) British botanist, who studied barley genetic resources.
Benyard (?) British fruit-grower.
Beria, Lavrentiy P. (?)
Berg, L.S. (1876-1950) Russian geographer and biologist; Academician of the USSR Academy of Sciences; from 1916 was a professor at Leningrad State University; in 1922-1929, Vice-Director of SIEA.
Bessey, C.E. (1845-1915) American botanist, phylogeneticist, Professor of Botany at the University of Nebraska.
Biffen, Rowlans (1874-1949) British plant breeder and geneticist; from 1908 to 1931, Professor of Cambridge University.
Blackman, V.H. (1872-1967) British biologist; from 1911 to 1937 Director of the Institute of Plant Physiology in London.

Boeuf, F. (?) French botanist; worked at the botanical gardens and an experimental station in Tunisia; from 1927 worked in the Museum of Natural History in Paris.
Borodin, D.N. (1889-?) Russian geobotanist, floriculturist and entomologist; from 1922 was in charge of the New York Branch of the Department of Applied Botany.
Borodin, I.P. (1847-1930) Russian botanist, professor of St. Petersburg University, 1899-1905 was the Head of Bureau of Applied Botany.
Brezhnev, D.D. (1905-1982) Russian plant scientist and geneticist, expert in tomato; in 1937-1941 was in charge of the Department of Vegetable Crops at VIR; in 1965-1978 took directorship of VIR.
Bridges, K. (1889-1938) American biologist and geneticist; in 1915-1938 worked at the Carnegie Institute in Washington. He participated in T.Morgan's research activities.
Budin, K.Z. (?) Russian agronomist, crop scientist, expert in potato; Academician of VASKhNIL; Head of Department of Potato VIR.
Bukasov, S.M. (1891-1983) Russian botanist and plant breeder, one of the most prominent experts in potato; Academician of VASKhNIL; from 1918 worked at the Department of Applied Botany later was Head of Potato Department, VIR.
Bukinich, D.D. (1882-1939) Russian irrigation engineer, geographer, ethnographer, archaeologist and traveller.
Burbank, L. (1849-1926) prominent American fruit breeder, founder (1875) of a horticultural and ornamental plant nursery in Santa Rosa, California.
Carleton, M.A. (1866-1925) American phytopathologist, worked at Bureau of Plant Industry USDA. He visited Russia and the Soviet Union several times.
Cavanilles, A.J. (1745-1804) Spanish botanist, Director of Royal Botanical Garden in Madrid.
Chernyakovskaya, E.G. (?) Russian botanist, who worked at the Botanical Gardens in Leningrad.
Chevalier, A.J.B. (1873-1956) French botanist, expert in tropical flora of Africa; Director of the Institute of Applied Botany in Paris.
Cook, O.F. (1867-1949) American biologist and plant breeder, head of the cotton department of the Bureau of Plant Industry, USDA.
Correns, K.E. (1864-1933) German botanist and geneticist; Director (1914-1933) of the Kaiser Wilhelm Biological Institute in Dahlem, Berlin. He rediscovered Mendel's laws simultaneously with Tschermak and De Vries.
Crew, F.A.E. (1888-1973) British geneticist, Director of the Institute of Animal Genetics in Edinburgh, member of the Agricultural Council in Scotland under the British Ministry of Agriculture.
Darlington, C.D. (1903 1981) British cytologist; from 1939 was the Director of the John Innes Institute, UK, later Professor of Botany at Oxford.
Darwin, Charles (1809-1882) prominent British biologist, discoverer of evolutionary laws.
Davenport, C.B. (1866-1944) American geneticist and eugenicist. In 1936 send a petition to the U.S. Department of State requesting public protest in defence of Soviet geneticists persecuted in the USSR.
De Candolle, A. (1806-1893) Swiss botanist, one of the founders of scientific plant geography.
De Vries, H. (1848-1935) Dutch botanist and geneticist, author of the theory of mutations. He rediscovered Mendel's laws simultaneously with Tschermak and Correns.
Dickon, W. (?) Cereal biologist, USDA.
Dickson, J. (?) American biologist, expert in plant protection problems.
Diels, L. (1874-1945) German botanist, plant taxonomist, morphologist and geographer.
Dobzhansky, F.G. (1900-1975) American geneticist of Russian origin; in 1921 graduated the

University in Kiev; in 1924-1927 worked in Leningrad University; in 1928 emigrated to the United States.

Dorofeev, V.F. (1919-1987) Russian agronomist, botanist, expert in wheat; Academician of VASKhNIL, was Head of the Department of Wheat, in 1978-1987 was the Director of VIR.

Dorset (?) American plant scientist, who worked at the Bureau of Plant Introduction, USDA.

Dragavtsev, V.A. Russian agronomist, geneticist; corresponding member of RASKhN; from 1991 the Director of VIR.

Dubinin, N.P. (?) Russian geneticist, Academician of the USSR Academy of Sciences; in 1932-1948 was in charge of the Department of Genetics in the Institute of Experimental Biology under the USSR Academy of Sciences.

Ducellier, L.O. (1878-1937) French botanist and plant breeder, Director of an experiment station in Algeria; professor at the University of Algeria.

Dumpis, M.F. (?) General Consul of the USSR in Afghanistan (1920s).

Eikhfeld, I.G. (1893-1989) Russian biologist and plant breeder, corresponding member of the USSR Academy of Sciences and Academician of VASKhNIL; in 1923-1940 was in charge of the Polar Branch of VIR; in 1940-1951 was the Director of VIR.

Engler, A.H.G. (1844-1930) German botanist and traveller, member of the Bavarian and Parisian Academies.

Enken, V.B. (1900-1981) Russian agronomist, botanist, expert in soya; from 1925 worked at the Institute of Applied Botany.

Enlow, C.R. (?) American plant scientist, who worked in the Section of Plant Introduction of the Bureau of Plant Industry, USDA.

Fairchild, D.G. (1869-1954) American agricultural researcher and traveller; founder and first director (1904-1928) of the Bureau of Plant Industry, USDA.

Famintsin, A.S. (1835-1918) Russian botanist and plant physiologist; Academician of the Russian Academy of Sciences.

Federley, H. (1879-1951) Finnish geneticist and plant breeder, from 1923 was professor at Helsinki University.

Fedotov, V.S. (?) Russian agronomist; researcher of Pushkin Department of VIR.

Filipchenko, Yu.A. (1882-1930) Russian zoologist and geneticist; founder of the Department of Genetics at St.Petersburg University and of the Laboratory of Genetics at the USSR Academy of Sciences.

Flyaksberger, Konstantin (1880-1939) Russian botanist, expert in cereal crop taxonomy; from 1907 worked with the Bureau of Applied Botany.

Fortunatova, O.K. (1898-1941) Russian agronomist. She worked with the Department of Applied Botany from 1923.

Frankel, O.H. (1900-1999) Austrian/Australian plant breeder; former head of CSIRO Plant Industry; was one of the first people to recognize the importance of landraces for plant breeding programmes. He played a major role in raising international awareness of the urgency of developing coherent strategies for the *ex situ* conservation of plant genetic resources.

Freyman, A.Yu. (1895-1959) Russian agronomist. She worked with the Department of Applied Botany from 1917, and was in charge of Moscow Division of this Department from 1922.

Fruwirth, H.C. (1862-1930) Austrian botanist, co-author and editor of a plant breeding manual.

Goldschmidt, R. (1878-1958) German biologist and geneticist; from 1935 was the Director of the Kaiser Wilhelm Biological Institute in Dahlem, Berlin; professor at the University of

California, U.S.A.

Golitzyn, Prince B.B. (1862-1916) Russian prince, statesman, from 1907 was a chairman of the Scientific Committe of the Ministry of Land Cultivation of Czarist Russia.

Gorbunov, N.P. (1892-1938) Soviet public figure; scientist, Academician of the USSR Academy of Sciences; private secretary of V.I. Lenin. He rendered a good deal of useful assistance to N.I. Vavilov in his efforts to organize the work of his Institute.

Govorov, L.I. (1885-1941) Russian agronomist and plant breeder; from 1915 worked at Moscow Breeding Station; from 1923 was the Head of the Steppe Station, and later was in charge of the Department of Leguminous Crops at VIR.

Grossgeim, A.A. (1888-1948) Russian botanist; Academician of the USSR Academy of Sciences; in 1917-1930 worked at the Polytechnical Institute in Tiflis (Tbilisi).

Gudzoni (?) A student of Rome University in the 1920s.

Gustafsson, O. (1908-1988) Swedish geneticist, worked 1947-1968 at the Swedish Forestry Research Institute and 1968-1974 at the Department of Genetics of Lund University.

Gustavson, O. (?) Swedish geneticist and plant breeder.

Haeckel, E.H. (1834-1919) German biologist, evolutionist, follower of Darwin, professor of zoology at the University of Jena.

Haldane, J.B.S. (1892-1964) British geneticist; from 1927 was the Head of Genetical Research of the John Innes Horticultural Institute.

Hall, Alfred Daniel (1864-1948) British botanist; Director of Rothamsted Experimental Station; from 1926, Director of the John Innes Horticultural Institute.

Hansen, N.E. (1866-1950) American agronomist and plant breeder, expert in forage and fruit crops. He visited Russia and the Soviet Union several times (1896-1908, 1913 and 1934).

Harlan, H.V. (1882-1944) American agronomist and expert in barley; worked at the Bureau of Plant Industry, USDA. He was a friend of N.I. Vavilov.

Harland, J. (?) British geneticist and plant breeder, Director of the British Experimental Station for cotton research and production in the Isle of Trinidad.

Hawkes J.G. (1915) British botanist, taxonomist, world expert in potato, University of Cambridge, later Professor of Botany at the University of Birmingham.

Hehn, V. (1813-1890) German botanist, who worked as a professor at the University of Derpt (Tartu); expert in cultivated plants.

Henry, A. (1857-1930) British botanist, plant collector. He collected for Kew for nearly 20 years and sent over 15 000 specimens, of which about 500 were new species, he wrote his great book, *Trees of Great Britain and Ireland*. He met N.I.Vavilov in Uruguay.

Ikeno, S. (1866-1943) Japanese botanist, geneticist, phylogeneticist; expert in barley, professor of botany at the University of Tokyo.

Ivanov, A.I. (1934-1988) Russian agronomist, plant scientist, expert in alfalfa; was Head of the Department of Forage Grasses of VIR.

Ivanov, D.S. (?) Russian agronomist, expert in rice; researcher at VIR.

Ivanov, N.N. (1884-1940) Russian chemist, biochemist and plant physiologist; from 1923 was a professor of Leningrad State University; from 1922 was in charge of the Biochemistry Laboratory of the Department of Applied Botany.

Ivanov, N.R. (1902-1978) Russian plant scientist and plant breeder; from 1926 worked at the Institute of Applied Botany, was the Institute's Director during the siege of Leningrad; from 1967 was the Scientific Secretary of the Commission on N.I. Vavilov's Heritage under the USSR Academy of Sciences; Vavilov prize winner.

Janossy, A. (1908-1975) Hungarian agronomist and plant breeder; founder and first Director of the Institute of Agrobotany in Tapiosele.

Johannsen, W.L. (1857-1927) Danish biologist; one of the fathers of genetics; in 1920s was the Director of the Institute of Plant Physiology at Copenhagen University.

Jones, L.R. (1864-1945) American biologist and plant protection specialist; collaborator of Bureau of Plant Industry of USDA, Professor at the University of Wisconsin.

Jones, Q. (?) American botanist, worked at the Bureau of Plant Industry, USDA, participated in several USDA plant explorations, in 1965 visited the USSR. He was the 1978 recipient of the Meyer Medal.

Juzepchuk, S.V. (1893-1959) Russian botanist, plant taxonomist and plant geographer; worked at Komarou Botanical Institute.

Kameraz, A.Y. (1904-1994) Russian agronomist, plant scientist and breeder, expert in potato; from 1927 worked at the Institute of Applied Botany.

Karpechenko, G.D. (1899-1942) Russian cytogeneticist; from 1925 was in charge of the Laboratory of Genetics at the Institute of Applied Botany.

Kasnowski, L. (?) Polish plant breeder.

Kato (?) Japanese botanist; Dean of Nanking University.

Keller, W. (?) German botanist and traveller.

Kerkis, Yu.Ya. (?) geneticist; in 1930s worked in the Institute of Genetics of the USSR Academy of Sciences.

Khoroshaylov, N.G. (1904-1995) Russian agronomist and crop scientist, expert in seed testing and production, from 1932 worked at VIR.

Kichunov, N.I. (1863-1942) Russian horticulturist; from 1921 was a professor and the Head of the Horticultural Department at Petrograd Agricultural Institute; from 1922 was employed as a research scientist at the Department of Applied Botany.

Kihara, H. (1893-1986) Japanese botanist, plant geneticist and cytologist; in 1920-1955 worked at Kyoto University, from 1955 was in charge the National Institute of Genetics.

Klokov, P.T. (?) Russian agronomist and chemist; professor of Moscow Zootechnical Institute.

Kobelev, V.K. (?) Russian plant scientist and breeder; worked at Turkestan Breeding Station and Central Asian Branch of VIR.

Kobylyansky, V.D. (?) Russian agronomist, botanist and geneticist, expert in rye; Head of the Department of Cereals of VIR.

Koelreuter, I.G.F. (1733-1806) German botanist, who in 1755-1761 worked in St. Petersburg, Russia. He was one of the founders of artificial hybridization techniques for plants.

Koernicke, F.A. (1808-1908) German botanist and taxonomist; curator of the Royal Herbarium in Berlin; later worked in the Botanical Gardens in St. Petersburg.

Koltsov, N.K. (1872-1940) Russian biologist and geneticist; corresponding member of the USSR Academy of Sciences; founder of the Institute of Experimental Biology; in 1918 started his professorship at Moscow State University.

Konarev, V.G. (?) Russian biochemist; Academician of VASKhNIL, Head of the Department of Molecular Biology of VIR.

Korzhinsky, S.I. (1861-1900) Russian botanist, geographer and biologist; Academician of the Russian Academy of Sciences; from 1892 was the chief botanist of St.Petersburg Botanical Gardens.

Kostov, D.S. (1897-1949) Bulgarian geneticist; member of the Bulgarian Academy of Sciences; from 1924 worked at Sofia and Harvard Universities; in 1939-1947 was the Director of the Central Agricultural Research Institute in Sofia.

Kovalev, N.V. (1888-1973) Russian agricultural botanist and crop expert; in 1928-1930 Director of Nikita Botanical Gardens, Crimea; afterwards Director of the All-Union Institute of Plant Protection; from 1931 was the Head of the Fruit Department at VIR and the Director of Maikop Experimental Station of VIR.

Krasochkin, V.T. (1904-1982) Russian agronomist and botanist, expert in beets; from 1926 worked at the Institute of Applied Botany.

Krasovskaya, I.V. (1896-1956) Russian plant physiologist. From 1925 to 1936 she worked at the Institute of Applied Botany.

Kreier, G.K. (1887-1942) Russian botanical geographer; from 1926 was in charge of the Section of Medicinal Plants at the Institute of Applied Botany.

Krivchenko, V.I. (?) Russian phytopatologist, expert in smut diseases; Head of the Department of Immunity, in 1987-1991 was the Director of VIR.

Kuckuck, H. (?) German botanist and barley expert.

Kuleshov, N.N. (1890-1968) Russian crop scientist, ecologist and taxonomist; full member of the Ukrainian Academy of Sciences; in 1915-1920 worked at Kharkov Agricultural Experimental Station, Ukraine; from 1925 was a research scientist of the Institute of Applied Botany.

Kuptsov, A.I. (1900-1986) Russian plant breeder and agrobotanist; in 1928-1933 worked at the Institute of Applied Botany.

Kuzmin, V. (?) Russian agronomist; research scientist of VIR.

Kuznetsov, V.A. (1877-1940) Russian botanist; in 1911 started his work with the Bureau; later was appointed Head of the Section of Forage Grasses.

La Gasca (?) Spanish botanist; in the beginning of the 19th century was the director of the Royal Botanic Gardens in Madrid; in 1818 set up a herbarium of cultivated plants.

La Gasca, M. (1776-1839) Spanish botanist; at the beginning of the 19th century was the director of the Royal Botanical Gardens in Madrid; in 1818 set up a herbarium of cultivated plants.

Lebedev, V.N. (1899-1957) Russian agronomist and plant breeder; was in charge of the department of cereals at Belotserkovskaya Station of the Sugar Trust, Ukraine.

Lehmann, Chr.O. (1926-1992) German plant scientist; was Head of German Genebank in Gatersleben.

Lekhnovich, V.S. (1902-1989) Russian plant scientist, expert in potato; from 1927 was researcher at the Institute of Applied Botany.

Lemeshev, N.K. (1938-1985) Russian geneticist and botanist, expert in cotton; was Head of the Department of Oil and Fiber Crops of VIR.

Lepin, T.K. (1895-1964) geneticist; worked at the Institute of Genetics. Levitsky, G.A. (1878-1942) Russian botanist and cytologist; corresponding member of the USSR Academy of Sciences; before 1925 was a professor of Kiev Polytechnical Institute; after 1925 was in charge of the Cytological Laboratory at the Institute of Applied Botany.

Likhonos, F.D. (1897-1984) Russian plant breeder and fruit-grower; in 1917-1925 worked at Voronezh Agricultural Institute; in 1925-1935 and, after an interval, from 1959 was a researcher in the Department of Fruit Plants at VIR.

Litvinov, Nikolai I. (?) Russian botanist; in 1908 became a scientific expert of the Bureau; in 1909-1917 was in charge of the Bureau's department in Voronezh.

Luss, Ya.Ya. (1897-1979) Russian geneticist and plant breeder, who worked in the field of fruit horticulture; in 1928- 1930 was a researcher at the Institute of Applied Botany.

Makasheva, R.H. Russian agronomist and crop scientist, expert in peas; was Head of the Department of Legume Crops of VIR.

Maltzev, A.I. (1879-1948) Russian scientist, expert in weedy plants; worked with the Bureau from 1908; in 1918-1926 was in charge of the Bureau's Steppe Zone Station.

Mansfeld, R. (1901-1960) German botanist and taxonomist; from 1949 was the first Director of German Genebank in Getersleben.

Markovich, V.V. (1865-1942) Russian silviculturist and floriculturist, expert in tropical plants; from 1925 was a scientific expert of the Institute of Applied Botany.

Mashtaler, G.A. (?) Russian expert in genetics, one of Lysenko's followers.

Matsuura, H. (?) Japanese geneticist, professor at the University of Hokkaido.

Maximov, N.A. (1880-1952) Russian plant physiologist, full member of the USSR Academy of Sciences; in 1917 became a professor at Tiflis (Tbilisi) Polytechnical Institute; from 1922 worked at the Department of Applied Botany.

Meister, G.K. (1873-1943) Russian plant breeder and geneticist; from 1918 was the Head of the Breeding Department at Saratov Agricultural Experimental Station; in 1920-1935 was the Station's Director.

Mendel, G. (1822-1884) prominent Austrian biologist and geneticist, discoverer of laws of heredity.

Merezhko, A.F. (?) Russian agronomist and geneticist, expert in wheat; Head of the Department of Wheat of VIR.

Meshchaninov, I.I. (1883-1967) Russian archaeologist and linguist, Academician of the USSR Academy of Sciences.

Meyer, F. (1875-1918) American prominent plant collector, botanist and horticulturist. He visited Russia and the Soviet Union several times.

Michurin, I.V. (1855-1936) Russian biologist and fruit breeder; Academician of VASKhNIL. In 1928 he founded a fruit plant nursery, which was in 1935 reorganized into the Central Genetic Laboratory of VASKhNIL.

Morrison, B.Y. (?-1965) American plant breeder, expert in azaleas, Director of the National Arboretum, Director of the USDA Bureau of Plant Industry, 1934 to 1949.

Morse, W.J. (1872-1931) American agronomist and botanist; in 1907-1929 worked at the USDA Bureau of Plant Industry.

Mosley (?) English biologist

Muller, H.J. (1890-1967) American geneticist, one of the founders of radiation genetics; Nobel Prize winner of 1946.

Muralov, N.I. (1886-1937) Soviet public official; in 1928 became the People's Commissar for Agriculture of the RSFSR; in 1933-1936 was Deputy Commissar for Agriculture of the USSR; in 1935-1937 President of VASKhNIL.

Navashin, S.G. (1857-1930) Russian cytologist and plant embryologist, Academician of the USSR Academy of Sciences, professor of the University of Kiev (1894-1915) and Tbilisi (1918- 1923); in 1923-1929 was the Director of the Timiryazev Biological Institute in Moscow.

Nilsson-Ehle, G. (1873-1949) Swedish biologist and geneticist, foreign corresponding member of the USSR Academy of Sciences, director of Svalof Breeding Station.

Oberg, E. (1909-1983) Swedish agronomist and evolutionist, he was professor of crop husbandry at the Swedish College of Agriculture, which 1978 became the Swedish University of Agricultural Sciences. In 1937 he visited the Soviet Union.

Offerman (?) Argentine geneticist, Dr Muller's assistant, who worked at the Institute of Genetics of the USSR Academy of Sciences.

Pashkevich, V.V. (1881-1939) Russian biologist and fruit-grower; Academician of VASKhNIL; from 1922 in charge of the fruit section of the Department of Applied Botany; professor of Leningrad Agricultural Institute.

Percival, J. (1863-1949) British botanist and wheat specialist; Professor of the University College of Reading, UK.

Pieters (?) American expert in forage grasses, researcher of the Division of Forage Crops and Diseases at the USDA Bureau of Plant Industry.

Pisarev, V.E. (1883-1972) Russian agronomist, crop breeder and geographer; founder of Tulun Experimental Station; from 1921 scientific expert of the Department of Applied Botany; from 1925 Deputy Director of Vavilov.

Popov, M.G. (1893-1955) Russian botanist and taxonomist; corresponding member of the

Ukrainian Academy of Sciences; in 1917-1921 lectured at Saratov University; in 1921-1927 professor of the Central Asian University.

Popovsky, M. (?) Russian writer and journalist; author of a number of books about Russian scientists.

Prezent, I.I. (1902-1968) Russian biologist; Academician of VASKhNIL; professor of Leningrad State University; leading ideologist of Lysenko's conceptual views.

Pryanishnikov, D.N. (1865-1948) Russian botanist, physiologist and agrochemist; Academician of the USSR Academy of Sciences; from 1895 professor of Petrovskaya Agricultural Academy; one of Vavilov's teachers.

Przhevalsky, N.M. (1839-1888) Russian geographer, explorer of the Central Asia; honoured member of the Russian Academy of Sciences.

Punnett, R.C. (1875-1967) British geneticist, professor of genetics at Cambridge University.

Razumov, V.I. (?) Russian plant physiologist, researcher of the Department of Genetics at VIR.

Regel, Robert E. (1867-1920) Russian botanist, 1905-1920 was the Head of Bureau of Applied Botany.

Reznik, S.E Russian writer, N.I.Vavilov's biographer.

Rodina, L.M. (?) Russian agronomist, expert in oats; worked at VIR.

Rozanova, M.A. (1885-1957) Russian botanist. From 1925 she worked at the Institute of Applied Botany.

Rudzinsky, D.L. (1866-1954) Russian plant breeder; from 1898 assistant, and later professor of Moscow Agricultural Institute, where in 1909 he organized Moscow Breeding Station one of the first plant breeding stations in Russia. In 1922 he moved to Lithuania and founded the Dotnuva Breeding Station.

Russel, E.J. (1872-1965) British botanist, agriculturist and geographer, professor at London University.

Ryerson, K.A. (1892-1990) American horticulturist, In the 1930s he was the Director of the USDA Bureau of Plant Industry. Dean, Emeritus, College of Agriculture, University of California, Davis (1937-1952), former Head of the Office of Foreign Plant Introduction, and former Chief of the USDA Bureau of Plant Industry. He was the 1968 recipient of the Meyer Medal.

Saks (?) Russian plant scientist, researcher of the Department of Genetics and Plant Breeding at VIR.

Sapegin, A.A. (1883-1946) Russian botanist and cytologist, Academician of the Ukrainian Academy of Sciences, director of Odessa Agricultural Experimental Station.

Satsyperov, F. (?) Russian botanist; in 1908 became a scientific expert in sunflower of the Bureau of Applied Botany.

Saunders, W. (1867-?) Canadian agronomist; professor of chemistry in the Central University of Canada; chief expert in cereals.

Schull, G.H. (1876-1954) American geneticist and plant breeder. In 1915-1942 was a professor of botany and genetics of Princeton University.

Schulz, A.A. (1855-1922) Russian agronomist, Vice Chairman of the Agricultural Scientific Committee.

Scott, D.H. (1854-1934) British paleobotanist, plant evolutionist, follower of Darwin.

Serebrovsky, A.S. (1892-1948) Russian biologist, Corresponding Member of the USSR Academy of Sciences, Academician of VASKhNIL. In 1920's was a professor of Moscow Zootechnical Institute.

Shevchuk, T.N. (?) Russian agronomist and crop scientist; was a researcher at VIR.

Shlykov, G.N. (1903-1977) Russian plant scientist; from 1931 was a researcher at VIR; worked at the Subtropical Department of this Institute and held the position of the

Institute's Vice Director. He was one of those who supplied information against Nikolai Vavilov.

Shmuk, A.A. (1886-1945) Russian agrochemist, soil and plant biochemist; Academician of VASKhNIL; professor of Kuban Agricultural Institute.

Sinskaya, E.N. (1889-1965) Russian agronomist, botanist and geographer. In 1921 she started her work at the Department of Applied Botany.

Sizov, I.A. (1900-1968) Russian plant breeder, head of the VIR's Department of Genetics and Plant Breeding, in 1961-1962 was the Director of VIR.

Skrdla, W. (?) American agronomist; was in charge of Northern Central PI Station of USDA in Ames, Iowa.

Stoletova, E.A. (1889-1964) Russian agronomist and plant breeder. In 1918 she started working at the Department of Applied Botany.

Stubbe, H. (1902-1989) German geneticist; from 1943 was founder and first Director of the Institut für Kulturpflanzenforschung in Gatersleben.

Sveshnikova (?) Russian geneticist. She worked at the Institute of Genetics under the USSR Academy of Sciences.

Swingle, W.C. (1875-1952) American agronomist and botanist. In 1891-1941 worked with the USDA Bureau of Plant Industry. He participated in a number of missions collecting subtropical plants.

Talanov, V.V. (1871-1936) Russian plant breeder and seed production specialist; Corresponding Member of the USSR Academy of Sciences. In 1917-1922 was a professor at Omsk Agricultural Institute and the director of a breeding station. In 1925 started his work as Vavilov's Vice Director.

Tanaka, T. (?) Japanese biologist, expert in citrus plants, professor of Taiwan University, organizer of an experimental station in Taiwan.

Ter-Avanesyan, D.V. (1909-1979) Russian agronomist, expert in cotton; from 1940 worked at VIR, was in charge of different Departments of VIR.

Thellung, A. (?) professor of botany in Zürich.

Timiryazev, K.A. (1843-1920) Russian naturalist and plant physiologist, expert in photosynthesis; Corresponding Member of the Russian Academy of Sciences.

Timofeyev-Resovsky, N.V. (1900-1981) Russian geneticist; one of the fathers of radiation genetics, biogeocoenology and molecular biology. In 1925-1945 worked at the Institute of Biology in Berlin.

Trabut, L. (?) French botanist; professor of medicinal botany at the Medical Department of the Algerian University.

Trofimovskaya, A.Ya. (1903-1991) Russian plant scientist, expert in barley. From 1940 she was a doctoral student and then a researcher of VIR, later was in charge of the Department of Cereals of VIR.

Tulaikov, N.M. (1875-1938) Russian agronomist and soil scientist; Academician of the USSR Academy of Sciences. In 1910-1916 was the organizer and director of Bezenchuk Agricultural Experimental Station. In 1920 started his professorship at Saratov Agricultural Institute.

Tumanyan, M.G. (1886-1950) Armenian agronomist and plant breeder, professor of Yerevan Agricultural Institute.

Tupitsyn, D.I. (?) Russian botanist, expert in plums; was in charge of Maikop Experimental Station of VIR.

Udachin, R.A. (?) Russian agronomist and botanist, expert in wheat; was a researcher at VIR.

Valle, O. (1899-1965) Finnish geneticist and plant breeder, expert in clover, was professor of crop husbandry and headed the Finnish Agricultural Research Centre in Dickursby.

Vavilov, I.I. (1859-1928) Russian merchant, N.I.Vavilov's father.
Vavilov, S.I. (1891-1951) Russian physicist, 1945-1951 was the President of Academy of Science of the USSR, N.I.Vavilov's brother.
Vavilov, Yu.N. (?) Russian scientist, Professor of Physics; N.I.Vavilov's son.
Vavilova, A.I. (1886-1940) Russian physician, N.I.Vavilov's sister.
Vavilova, L.I. (1893-1914) Russian microbiologist, N.I.Vavilov's sister.
Vilmorin, J. (1882-1933) French agronomist; in 1917-1933 was the head of Vilmorin-Andrieux, the family business company dealing in plant breeding and seed production.
Voronov, Yu.N. (1874-1931) Russian geobotanist and taxonomist, expert in subtropical plant diversity. In 1918-1921 was the director of Tiflis Botanical Gradens. In 1925 became a research scientist of the Institute of Applied Botany.
Voskresenskaya, O.A. (1904-1949) Russian agronomist, expert in potato; from 1930 was a researcher at VIR.
Vysotsky, G.N. (1865-1940) Russian soil scientist, geobotanist and silviculturist; Academician of the Ukrainian Academy of Sciences and VASKhNIL; in 1920 was a professor at several agricultural institutes.
Westover, H.L. (?) American botanist; worked at USDA.
Whitehouse, W.E. (1893-1982) American botanist; worked at USDA, participated in several USDA plant explorations, in 1929 he collected the pistachio seed in Persia that gave rise to the US pistachio industry. He visited Russia and the Soviet Union several times. He was the 1968 recipient of the Meyer Medal.
Yachevsky, A.A. (1863-1932) Russian micologist and phytopathologist; Corresponding Member of the USSR Academy of Sciences; from 1917 was in charge of the Department of Micology and Phytopathology in the framework of the Agricultural Scientific Committee.
Yakovlev, Ya.A. (1896-1938) Soviet public and communist party leader; from 1929 was the USSR People's Commissar for Agriculture; from 1934 was in charge of the Agricultural Sector in the Central Committee of the Communist Party.
Yakushevsky, E.S. (1902-1989) Russian plant scientist and breeder; worked at VIR.
Yakushina, O.V. (?) Russian plant scientist. She worked as an assistant professor at the Department of Special Agriculture of Saratov Agricultural Institute.
Zade, A. (?) German phytopathologist and plant breeder; professor of the Agronomy Institute at Leipzig University.
Zaitsev, G.S. (1887-1929) Russian plant scientist and breeder; in 1925 became the head of breeding divisions of a number of experimental stations. From 1925 was in charge of the cotton research section of the Institute of Applied Botany.
Zegalov, S.I. (1881-1927) Russian geneticist, plant breeder and plant scientist. From 1920 worked at the Timiryazev Agricultural Academy as a professor and from 1922 as the head of its Moscow Experimental Station.
Zhukovsky, P.M. (1888-1975) Russian botanist, Academician of VASKhNIL; in 1915-1925 worked at Tiflis Botanical Gardens; from 1925 was a research scientist of the Institute of Applied Botany; in 1951-1965 was appointed Director of VIR.

Appendix III. Vavilov's major collecting expeditions, 1916-1940

1916 expedition to Iran (Hamadan and Khorasan) and Pamir (Shungan, Rushan and Khorog)

1921 acquaintance visit to Canada (Ontario) and USA (New York, Pennsylvania, Maryland, Virginia, North and South Carolina, Kentucky, Indiana, Illinois, Iowa, Wisconsin, Minnesota, North and South Dakota, Wyoming, Colorado, Arizona, California, Oregon, Maine)

1924 expedition to Afghanistan (Herat, Afghan Turkestan, Gaimag, Bamian, Hindu Kush, Badakhshan, Kafiristan, Jalalabad, Kabul, Herat, Kandahar, Baquia, Helmand, Farakh, Sehistan), accompanied by D.D. Bukinich and V.N. Lebedev

1925 expedition to Khoresm (Khiva, Novyi Urgench, Gurlen, Tashauz)

1926-1927 expedition to Mediterranean countries (France, Syria, Palestine, Transjordan, Algeria, Morocco, Tunisia, Greece, Sicily, Sardinia, Cyprus, Crete, Italy, Spain, Portugal and Egypt where Gudzoni was explored by Vavilov's request) and to Abyssinia (Djibouti, Addis Ababa, banks of Nile, Tsana Lake), Eritrea (Massaua) and Yemen (Hodeida, Jidda, Hedjas)

1927 exploration of mountainous regions in Wuertemberg (Bavaria, Germany)

1929 expedition to China (Xinjiang – Kashgar, Uch-Turfan, Aksu, Kucha, Urumchi, Kulja, Yarkand, Hotan) together with M.G. Popov, then alone to China (Taiwan), Japan (Honshu, Kyushu and Hokkaido) and Korea.

1930 expedition to USA (Florida, Louisiana, Arizona, Texas, California), Mexico, Guatemala and Honduras

1932-1933 visit to Canada (Ontario, Manitoba, Saskatchewan, Alberta, British Columbia) and USA (Washington, Colorado, Montana, Kansas, Idaho, Louisiana, Arkansas, Arizona, California, Nebraska, Nevada, New Mexico, North and South Dakotas, Oklahoma, Oregon, Texas, Utah)

1932–1933 expedition to Cuba, Mexico (Yucatan), Ecuador (Cordilleras), Peru (Lake Titicaca, Puno Mt., Cordilleras), Bolivia (Cordilleras), Chile (Panama River). Brazil (Rio de Janeiro, Amazon), Argentina, Uruguay, Trinidad and Puerto Rico

1921-1940 systematic explorations of the European part of Russia and the whole regions of the Caucasus and Middle Asia.

Appendix IV. Major collecting missions carried out by Vavilov's associates, 1922-1933

1922-1923	expedition of V.E. Pisarev and V.P. Kuzmin to Mongolia
1923	expedition of E.I. Barulina to Crimea (Ukraine)
1924	expedition of E.I. Sinskaya to Altai
1925-1926	expedition of S.M. Bukasov and Yu. N. Voronov to Mexico, Guatemala and Colombia
1925-1926	expedition of E.N. Stoletova to Armenia
1925-1927	expedition of P.M. Zhukovsky to Turkey
1926	expedition of N.N. Kuleshov and V.V. Pashkevich to Azerbaijan
1926	expedition of K.A. Flyaksberger to Azerbaijan and Russia (Daghestan)
1926	expeditions of N.N. Kuleshov and V.K. Kobelev to Uzbekistan
1926	expedition of K.A. Flyaksberger to Far East of Russia
1926-1928	expedition of V.V. Markovich to Palestine, Pakistan, India, Java and Ceylon
1926-1928	expedition of S.V. Yuzepchuk to Peru, Bolivia and Chile
1927	expedition N.N. Kuleshov to Turkmenia
1927	expedition K.G. Kreier to Central and Western part of Siberia
1928-1929	expedition of E.N. Sinskaya to Japan
1928-1932	expedition of G.K. Kreier to Georgia and Azerbaijan
1930	expedition of E.A. Stoletova to Georgia (USSR)
1930	expedition of G.K. Kreier to Kirgizia and Uzbekistan
1933	expedition of E.I. Barulina to Georgia (USSR)

Appendix V. VIR's major expeditions within the former USSR, 1986-1990

1986

European part of the USSR

Russia: Leningrad reg.; weedy plants; 17 accessions and 811 specimens of herbarium.
Russia: Kaliningrad reg.; fruits, berries and vegetables; 163 accessions.
Russia: Astrakhan reg., Kalmyk reg.; fruits and berries; 89 accessions.
Russia: Ryazan reg., Vladimir reg., Ivanov reg., Yaroslavl reg., Tver reg.; *Actinidia* spp. and *Schizandra* spp.; 90 accessions and 124 specimens of herbarium.
Ukraine; *Prunus* spp., *Armeniaca* spp., *Cerasus* spp., *Pyrus* spp. and *Malus* spp.; 88 accessions.
Ukraine: Crimea pen.; *Juglans* spp.; 15 accessions.
Ukraine, Moldavia; local cereals; 286 accessions.

Caucasus and Transcaucasus

Russia: Krasnodar reg, Stavropol reg.; forages and berries; 12 accessions.
Georgia: Abkhazia; vegetables, fruits and berries; 121 accessions.
Georgia; *Citrus* spp.; 51 accessions.
Georgia; *Lupinus* spp.; 11 accessions.
Georgia; *Malus* spp.; 28 accessions.
Georgia, Azerbaijan and Armenia; subtropical fruits and nuts; 110 accessions.
Azerbaijan; *Triticum* spp. and local cereals; 66 accessions.
Azerbaijan, Armenia; forages; 33 accessions and 2118 specimens of herbarium.

Central Asian republics

Uzbekistan, Tadjikistan; *Gossypium* spp.; 35 accessions.
Uzbekistan, Tadjikistan, Tukmenistan; fruits; 78 accessions.
Uzbekistan, Tadjikistan, Kirgizstan; fruits and vegetables; 144 accessions.
Turkmenistan; *Vitis* spp. and fruits; 104 accessions.

Kazakhstan

Kazakhstan; local cereals and forages; 36 accessions.
Kazakhstan; fruits; 87 accessions.

Western part of Siberia

Russia: Altai reg.; local cereals, vegetables and potato; 180 accessions.

Eastern part of Siberia

Russia: Buryatia, Chita reg.; forages; 128 accessions.

Far East of Russia

Russia: Amur reg.; vegetables, berries and forages; 32 accessions.
Russia: Khabarovsk reg.; *Lonicera* spp. and *Ribes* spp.; 56 accessions and 36 specimens of herbarium.
Russia: Primorski reg.; fruits and berries; 451 accessions.
Russia: Sakhalin is.; fruits and berries; 197 accessions.

1987

European part of the USSR

Russia: Karelia; forages; 180 specimens of herbarium.
Russia: Moscow reg.; vegetables; 67 accessions.
Ukraine; *Prunus* spp., *Malus* spp., *Cerasus* spp. and *Pyrus* spp.; 23 accessions.
Ukraine; forages; 107 accessions.

Caucasus and Transcaucasus

Russia: Krasnodar reg., Stavropol reg., Karachaevo-Cherkesia, Kabardino-Balkaria; forages and fruits; 2100 specimens of herbarium.
Russia: Krasnodar reg., Adygeya; *Sorghum* spp. and *Cicer* spp.; 40 accessions and 700 specimens of herbarium.
Russia: Krasnodar reg., Chechnya, Ingushetia, Osetia; Georgia; cereals, forages, vegetables and berries; 103 accessions and 700 specimens of herbarium.
Russia: Krasnodar reg., Stavropol reg., Daghestan, Chechnya, Ingushetia; fruits; 131 accessions.
Russia: Krasnodar reg.; Georgia, Azerbaijan and Armenia; fruits, berries and *Aegilops* spp.; 333 accessions.
Georgia: Abkhazia, Adzharia; subtropical fruits, berries and nuts; 176 accessions.
Georgia: Abkhazia, Adzharia; *Citrus* spp.; 116 accessions.
Georgia: Abkhazia, Adzharia; *Eucalyptus* spp., *Bambusa* spp. and *Aleurites* spp.; 29 accessions.

Central Asian republics

Tadjikistan, Uzbekistan; local cereals, legumes and vegetables; 53 accessions.
Turkmenistan; fruits and *Vitis* spp.; 71 accessions.
Turkmenistan; local cereals and forages; 1500 specimens of herbarium.

Western part of Siberia

Russia: Altai reg.; local cereals, forages and vegetables; 164 accessions.

Eastern part of Siberia

Russia: Buryatia; forages and berries; 58 accessions.
Russia: Yakutia; forages, legumes, vegetables and berries; 350 accessions and 151 specimens of herbarium.

Far East of Russia

Russia: Amur reg.; *Lonicera* spp., *Ribes* spp., *Schizandra* spp.; 114 accessions.
Russia: Kamchatka reg.; forages and berries; 80 accessions.
Russia: Primorski reg.; fruits and berries; 112 accessions.
Russia: Sakhalin is.; forages; 41 accessions.

1988

European part of the USSR

Russia: Karelia; forages, vegetables and berries; 25 accessions.
Russia: Volvograd reg., Rostov reg.; *Allium* spp. and *Capsicum* spp.; 26 accessions.
Belorussia, Lithuania, Latvia, Estonia; fruits and berries; 171 accessions.

Caucasus and Transcaucasus

Russia: Krasnodar reg., Chechya, Ingushetia, Osetia; Georgia; cereals, forages, vegetables

and berries; 97 accessions and 316 specimens of herbarium.
Russia: Stavropol reg., Daghestan; 110 accessions and 1000 specimens of herbarium.
Russia: Krasnodar reg.; forages; 48 accessions.
Georgia; vegetables and fruits; 70 accessions.

Middle Asia republics
Tadjikistan; *Amygdalus* spp., *Armeniaca* spp., *Cerasus* spp. and *Prunus* spp.; 151 accessions.
Turkmenistan; vegetables and forages; 56 accessions.
Turkmenistan; fruits, *Punica* spp., *Juglans* spp., *Amygdalus* spp. and *Vitis* spp.; 100 accessions.
Turkmenistan; *Ficus* spp.; 24 accessions and 12 specimens of herbarium.
Kirgizstan; fruits; 91 accessions.

Kazakhstan
Kazakhstan; tau-sagyz; 154 accessions.
Kazakhstan; galophit forages; 106 accessions.
Kazakhstan; *Aegilops* spp.; 44 accessions.

Eastern part of Siberia
Russia: Yakutia; forages and galophits; 69 accessions.
Russia: Yakutia; legumes, vegetables and berries; 160 accessions and 1489 specimens of herbarium.

Far East of Russia
Russia: Khabarovsk reg.; fruits and berries; 73 accessions.
Russia: Amur reg.; forages and fruits; 90 accessions and 117 specimens of herbarium.
Russia: Primorski reg.; local cereals, forages and vegetables; 118 accessions.
Russia: Primorski reg.; fruits, berries and *Vitis* spp.; 222 accessions.

1989

European part of the USSR
Russia: Arkhangelsk reg.; forages and vegetables; 94 accessions.
Russia: Novgorod reg., Vologda reg., Arkhangelsk reg.; forages, vegetables and fruits; 42 accessions and 1050 specimens of herbarium.
Russia: Saratov reg.; *Cerasus* spp., *Prunus* spp., *Malus* spp. and *Cydonia* spp.; 27 accessions.
Moldavia; fruits and berries; 180 accessions.

Caucasus and Transcaucasus
Russia; Krasnodar reg., Osetiya, Chechnya, Ingushetiya: Georgia: Abkhazia; forages, vegetables and fruits; 458 accessions.
Russia; Krasnodar reg.; Georgia: Abkhazia, Adzharia; *Citrus* spp.; 44 accessions.

Central Asian republics
Uzbekistan; vegetables, fruits and *Vitis* spp.; 91 accessions.
Uzbekistan; galophit forages; 70 accessions.
Uzbekistan; *Gossypium* spp.; 27 accessions.
Turkmenistan; fruits; 129 accessions and 400 specimens of herbarium.
Turkmenistan; *Punica* spp. and *Ficus* spp.; 36 accessions.
Turkmenistan; *Vitis* spp.; 11 accessions.

Kazakhstan

Kazakhstan; tau-sagyz and kok-sagyz; 133 accessions.

Eastern part of Siberia

Russia; Yakutia; forages, legumes, vegetables and fruits; 140 accessions.
Russia; Buryatia; forges, legumes and fruits; 50 accessions.

Far East of Russia

Russia; Khabarovsk reg.; forages; 85 accessions.
Russia; Amur reg.; berries; 53 accessions and 150 specimens of herbarium.
Russia; Primorski reg.; fruits; 34 accessions and 64 specimens of herbarium.
Russia; Primorski reg., Sakhalin is.; fruits and berries; 96 accessions.

1990

European part of the USSR

Russia: Leningrad reg., Pskov reg.; weeds and forages; 54 accessions and 2431 specimens of herbarium.
Russia: Vladimir reg., Ryazan reg., Moscow reg.; forages; 96 accessions.
Russia: Kalmykia, Astrakhan reg., Volgograd reg., Saratov reg.; *Malus* spp., *Pyrus* spp., *Prunus* spp., *Cerasus* spp., *Cydonia* spp.; 87 accessions.
Russia: Krasnodar reg.; legume forages; 88 accessions.
Belorussia; weeds and forages; 50 accessions and 2000 specimens of herbarium.
Ukraine; *Linum* spp.; 25 accessions and 1200 specimens of herbarium.
Ukraine: Crimea pen.; Moldavia; fruits; 217 accessions.

Caucasus and Transcaucasus

Russia: Krasnodar reg.; Georgia: Abkhazia; subtropical fruits, berries and nuts; 54 accessions.

Central Asian republics

Uzbekistan, Tadjikistan; fruits, vegetables and *Vitis* spp.; 52 accessions.
Turkmenistan; fruits; 74 accessions and 150 specimens of herbarium.
Turkmenistan; *Punica* spp.; 12 accessions.
Turkmenistan; *Vitis* spp.; 20 accessions.

Kazakhstan

Kazakhstan; *Haloxylon* spp.; 65 accessions.

Eastern part of Siberia

Russia: Krasnoyarsk reg.; legumes, forages, vegetables and berries; 201 accessions.
Russia: Yakutia; forages; 31 accessions.
Russia: Buryatia; forages and berries; 37 accessions.

Far East of Russia

Russia: Khabarovsk reg.; fruits and berries; 102 accessions
Russia: Amur reg.; *Ribes* spp.; 93 accessions.
Russia: Primorski reg.; *Malus* spp. and *Pyrus* spp.; 67 accessions.
Russia: Primorski reg.; vegetables; 164 accessions.
Russia: Sakhalin is.; fruits and berries; 32 accessions.

Appendix VI. VIR's major expeditions to foreign countries, 1954-1994

1954 France, Corsica; by P.M. Zhukovsky; cereals, corn, legumes, vegetables, forages, fruits, ornamental plants – 6811 accessions.

1955 Argentina; by P.M. Zhukovsky; cereals, legumes, forages, industrial crops, potato, vegetables crops – 376 accessions.

1956 China, central and western provinces; by N.R. Ivanov *et al.*; cereals, legumes, corn, industrial crops, vegetables, forages, fruits, subtropical crops – 2100 acc.

1956 Bulgaria, explored central, southern and mountain regions; by I.A. Sizov *et al.*; cereals, legumes and vegetables – 606 acc.

1956-1959 India, Nepal; by D.V. Ter-Avanesyan; cereals, legumes, vegetables, industrial and oil crops, forages, fruits, subtropical crops – 4300 acc.

1957 China; by D.I. Tupitsin, A.M.Gorskii; cereals, legumes, vegetables, forages, industrial and oil crops, ornamental, fruits, subtropical crops, grapes – 1117 acc.

1957-1959 Mongolia; by N.G. Khoroshailov; cereals, small grains – 287 acc.

1958 Argentina, Peru, Chile, Mexico; by P.M. Zhukovsky; cereals, corn, legumes, potato, vegetables, industrial crops, forages – 1214 acc.

1958-1959 Egypt, Ethiopia, Sudan; by F.F. Sidorov, T.Ya. Zarubailo; cereals, legumes, industrial crops, vegetables, tuber crops, forages, fruits, subtropical crops – 1109 acc.

1959 Iraq; by T.N. Shevchuk; cereals, legumes, vegetables, forages, industrial and oil crops, fruit crops – 314 acc.

1959-1960 Bulgaria; by G.I. Miroshnichenko; cereals, corn, legumes, vegetables and tuber crops – 2412 acc.

1963 Afghanistan, explored main agricultural regions; by T.N. Shevchuk, V.L. Vitkovsky; cereals, legumes, vegetables, forages, fruits, subtropical crops, grapes – 725 acc.

1967 Chile, main agricultural regions; by G.E. Shmaraev, A.G. Zykin; cereals, corn, vegetables, forages, potatoes – 2028 acc.

1967 Turkey; by V.F. Dorofeev; cereals, forages, vegetables, industrial and oil crops, ornamental plants – 510 acc.

1968 Sudan; by I.I. Miroshnichenko, V.P. Gorbunov; cereals, legumes, vegetables, forages, industrial and oil crops – 517 acc.

1968 Tanzania, Uganda; by G.E. Shmaraev, N.I. Korsakov; cereals, legumes, vegetables, forages, industrial and oil crops – 980 acc.

1968 Iran; by V.F. Dorofeev, V.L. Vitkovsky; cereals, legumes, vegetables, industrial and oil crops, fruits, subtropical crops – 2894 acc.

1968 Mexico, south states, mountain regions and volcano in Mexico State; by K.Z. Budin; cereals, legumes, vegetables, more 30 species of potato– 1176 acc.

1968 Brazil, western and coastal parts; by T.N. Shevchuk, A.G. Zykin; cereals, legumes, vegetables, tuber crops, forages, industrial and oil crops – 424 acc.

1968 Ethiopia, explored N.I. Vavilov's route; by V.M. Berlyand-Kozhevnikov; cereals, sorghum, legumes, vegetables, industrial and oil crops – 2730 acc.

1969 India, main agricultural regions; by V.F. Dorofeev, B.V. Shver; cereals, legumes, rice, vegetables, oil crops, fruits, grapes – 938 acc.

1969 Algeria; by K.Z. Budin, V.L. Vitkovsky; cereals, legumes, vegetables, forages, industrial and oil crops, fruits, subtropical crops – 840 acc.

1969 Afghanistan, explored N.I. Vavilov's route; by V.M. Berlyand-Kozhevnikov; cereals (*Triticum* spp., *Hordeum* spp., *Avena* spp.), legumes, vegetables, industrial crops, fruits plants – 1500 acc.

1970 Peru, south mountain parts – Lima, Uankaho, Cusco, Puno districts; by K.Z. Budin, G.E. Shmaraev; wheats, barley, corn, rice, cotton, cultivated and wild potato, tuber crops – 1327 acc.

1970 Tunisia, Morocco; by A.V. Pukhalsky, E.V. Mazhorov; wheats, barley, corn, vegetables, legumes, cotton, ornamental plants – 500 acc.

1970 Pakistan; by V.F. Dorofeev, N.I. Korsakov; cereals, legumes, forages, vegetable crops – 856 acc.

1971 Guinea, Mali, Senegal; by N.I. Korsakov, N.K. Lemeshev *et al.*; cereals, legumes, forages, vegetables, fruits and industrial crops – 2218 acc.

1971 Bolivia, departments of La-Paz, Cocabamba, Santa-Cruz and Potosi, collected 738 accessions; Peru, southwestern parts – departments of Cuzco, Lima and Guankaho, collected 403 accessions; Ecuador, western parts in mountain and coastal regions, collected 1,249 accessions; wheat, corn, barley, legumes, vegetables, 14 species of potato, industrial and fruit crops.

1972 Burundi, Kenya, Somalia; by V.F. Dorofeev, N.P. Agafonov; cereals, legumes, vegetables, forages, citrus and industrial crops – 799 acc.

1973 Argentina, north and northeastern provinces; Uruguay; by G.E. Shmaraev, R.A. Udachin *et al.*; cereals, corn, potato and industrial crops – 1600 acc.

1974 Peru; by N.K. Lemeshev, L.E. Gorbatenko; 2,600 accessions of potato; corn, rice, cotton, tomato, tobacco, pepper, sunflower – all in all 4918 acc.

1974 Nepal, by A.G. Lyakhovkin; cereals, legumes, vegetables and oil crops – 994 acc.

1974 Syria; Iraq, main agricultural regions; by V.F. Dorofeev; wheat, barley, legumes, vegetables and industrial crops – 778 acc.

1974 Egypt, Libya; by Ya.S. Nesterov, R.A. Udachin; cereals, legumes, vegetables, forages, citrus, industrial and fruit crops – 755 acc.

1974 Cameroon; by G.E. Shmaraev, H.I. Korsakov *et al.*; cereals, legumes, vegetables, forages, industrial and tuber crops – 746 acc.

1975 Mexico; by N.K. Lemeshev; cereals, legumes, vegetables, forages and industrial crops – 336 acc.

1975 Colombia, explored mountain regions; Venezuela, explored main agricultural regions; by G.E. Shmaraev, L.E. Gorbatenko; cereals, legumes, potato and industrial crops – 2130 acc.

1976 India; central areas; by V.L. Vitkovsky; cereals, fruits and *Citrus* spp. – 723 acc.

1977 Spain, agricultural regions following N.I. Vavilov's trip; by L.E. Gorbatenko; cereals, legumes, vegetables and fruits – 5074 acc.

1977 Portugal; by Ya.S. Nesterov *et al.*; cereals, legumes, vegetables and forages – 3045 acc.

1977 Korea (KPDR); by G.V. Eremin; cereals, vegetables and fruits – 147 acc.

1977 Philippines; by A.G. Lyakhovkin, V.N. Soldatov; cereals, legumes, vegetables and industrial crops – 1173 acc.

1977 Upper Volta; by S.N. Bakhareva, S.G. Varadinov; cereals, legumes, vegetables, forages, citrus, industrial and tuber crops – 846 acc.

1977 Ghana; by V.S. Sotchenko, G.A. Tekhanovich; cereals, legumes, vegetables, forages, industrial and fruit crops – 912 acc.

1978 Czechoslovakia, Slovakia, Moravia, Czehia; by E.V. Mazhorov, K.A. Kobylyanskaya; *Triticum* spp, *Hordeum* spp, *Avena* spp, legumes, forages and industrial crops – 1521 acc.

1978 Afghanistan, all fruit-growing regions; by V.P. Denisov *et al.*; fruits, berries and ornamental crops – 644 accessions and grafts.
1978 Yemen (YPDR), main agricultural regions; by V.S. Sotchenko, G.V. Podkuichenko *et al.*; cereals, legumes, vegetables and industrial crops – 297 acc.
1978 Algeria; by V.D. Kobylyanski, S.N. Bakhareva; cereals, legumes, vegetables, forages, industrial and fruit crops – 347 acc.
1978 Pakistan; by N.I. Korsakov, G.G. Davidyan *et al.*; cereals, legumes, vegetables, fruits and industrial crops – 1155 acc.
1978 Sudan; by B.Kh. Sattarov; cereals, legumes, vegetables and industrial crops – 109 acc.
1978 Trinidad and Tobago, Jamaica; by V.L. Vitkovsky, L.E. Gorbatenko; cereals, legumes, vegetables, potato, fruits and industrial crops – 399 acc.
1979 Bangladesh, main agricultural regions; by V.F. Dorofeev; rice, corn, legumes, vegetables and industrial crops – 400 acc.
1979 Malagasy Republic; by G.G. Davidyan, S.N. Bakhareva; cereals, legumes, forages, vegetables, fruits, tuber and industrial crops – 1555 acc.
1980 Bolivia, Cocabamba, Santa-Cruz, Potosi and Cukisaka; by A.G. Zykin; cereals, species of potato and sunflower – 649 acc.
1980 Burma, central and south parts; by K.A. Kobylyanskaya, V.F. Chapurin; cereals, legumes and vegetables – 201 acc.
1980 Peru, main agricultural regions of mountain, coastal and tropical parts; by A.F. Merezhko, G.A. Tekhanovich; cereals, legumes, potato species and cotton – 1100 acc.
1980 Korea (KPDR); by T.N. Ulyanova, V.I. Burenin; cereals, rice, corn, soya, vegetables – 201 acc.
1980 Gabon, Congo; by S.N. Bakhareva, S.G. Varadinov *et al.*; cereals, forages, fruits and industrial crops – 702 acc.
1980 Colombia; by Ya.S. Nesterov, V.A. Koshkin *et al.*; cereals, legumes, potato, vegetables and industrial crops – 1055 acc.
1981 Korea (KPDR); by T.N. Ulyanova, A.A. Yushev; cereals, legumes, vegetables, fruits and industrial crops – 246 accessions and 210 herbarium specimens.
1982 Canary and Balearic Islands; by V.I. Burenin, A.D. Barsukov; wild species of *Beta, Lupinus, Avena, Hordeum, Allium, Capsicum* – 558 acc.
1982 Zaire, Zambia; by S.N. Bakhareva *et al.*; cereals, legumes, vegetables, forages, fruits and industrial crops – 746 acc.
1982 Argentina, central regions; *Hordeum* spp, *Solanum* spp., *Helianthus* spp., *Gossipium* spp. – 1047 acc.
1983 Burundi, Tanzania; by S.N. Bakhareva, V.N. Soldatov *et al.*; cereals, legumes, vegetables, potato, fruits and industrial crops – 1294 acc.
1983 Laos; by K.A. Kobylyanskaya; cereals, legumes, vegetables, potato and industrial crops – 850 acc.
1984 Brazil, western and coastal regions; by N.K. Lemeshev, V.V. Gridneva; cereals, corn, legumes and industrial crops – 645 acc.
1984 Yemen (YAR), fruit-growing regions; by G.V. Podkuichenko, A.A. Yushev; cereals, legumes, forages, vegetables, mainly fruits, grapes and citrus – 549 accessions and grafts.
1984 Greece; by L.V. Semenova *et al.*; cereals, legumes, vegetables, forages and industrial crops – 750 acc.
1984 Nigeria; by A.I. Borodanenko *et al.*; cereals, legumes, vegetables and industrial crops – 80 acc.

1985 Turkey; by N.K. Lemeshev, I.G. Loskutov; cereals, legumes, forages, vegetables and oil crops – 206 acc.
1986 Sri Lanka; by N.P. Agafonov; cereals, legumes, forages, vegetables and fruits – 370 accessions and grafts.
1986 Botswana, Zimbabwe; by A.A. Yushev, G.A. Tekhanovich; cereals, legumes, vegetables, fruits and industrial crops – 1034 acc.
1986 Spain; by A.G. Zykin, N.K. Lemeshev *et al.*; cereals, legumes, forages, vegetables and industrial crops – 612 acc.
1986 Syria; by V.I. Burenin; cereals, legumes, vegetables and industrial crops – 370 acc.
1986 Benin; by S.N. Bakhareva, V.N. Soldatov; cereals, legumes, vegetables and industrial crops – 399 acc.
1987 Venezuela, explored 15 states from 20 ones; by L.E. Gorbatenko *et al.*; cereals, corn, legumes, *Gossypium* spp., *Solanum* spp. – 640 acc.
1987 Brazil; by V.I. Pyzhenkov, B.S. Kurlovich *et al.*; cereals, legumes, vegetables and industrial crops – 135 acc.
1987 Morocco; by V.N. Soldatov, R.A. Okcyuzyan; cereals, legumes, vegetables and industrial crops – 300 acc.
1988 Bolivia, explored departments of Kocabamba, La-Pas, Santa-Crus, Potosi and Cukisaka; by A.G. Zykin, L.E. Gorbatenko; cereals, legumes and vegetables – 920 acc.
1988 Peru, explored all agricultural regions; by N.K. Lemeshev, N.M. Zhitlova; a lot of forms potato, cotton, corn, bean and lupin - 400 acc.
1988 Nepal, explored lowland areas; by L.A. Burmistrov, I.G. Loskutov *et al.*; cereals, legumes, forages, fruits and citrus – 640 accessions and grafts.
1989 Bhutan, explored all territory; by G.E. Shmaraev, V.P. Denisov; cereals, legumes, vegetables and fruits – 500 accessions and grafts.
1989 Ecuador, explored mountain regions in eastern part; by B.S. Kurlovich, N.M. Vlasov; cereals, legumes, lupines, industrial crops, vegetables, species of potato, *Quinoa* and amaranth – 903 acc.
1989 Cote d'Ivoire; by S.N. Bakhareva, S.D. Kiru; cereals, legumes, forages, fruits, vegetables and industrial crops – 480 acc.
1990 Colombia, explored south regions; by L.A. Burmistrov, L.E. Gorbatenko; rice, beans, sorghum, peppers and potato – 1016 acc.
1990 Egypt, Canary islands, Guinea, South Africa, Namibia, Madagascar island, Seychelles, Indonesia, Singapore (China), visited by S.N. Bakhareva on the research ship; cereals, legumes, forages, vegetables, fruits, industrial and tuber crops – 562 acc.
1991 Egypt; by G.E. Shmaraev, A.A. Filatenko *et al.*; cereals, vegetables, industrial and oil crops – 1730 acc.
1991 Tunisia; by N.M. Zoteeva, S.D. Kiru; cereals, legumes, forages, vegetables and fruits – 540 acc.
1991 Portugal, explored north and south regions and Madeira island; by B.S. Kurlovich, N.B. Brach; corn, legumes, lupines, flax, vegetables and forages – 792 acc.
1991 Costa Rica; by N.K. Lemeshev, L.E. Gorbatenko; cereals, legumes, vegetables and industrial crops – 410 acc.
1994 Canada (last abroad expedition), visited research institutes; by N.K. Lemeshev, N.I. Dzyubenko; cereals, legumes, forages, vegetables, fruits and industrial crops – 805 acc.

Appendix VII. Joint collecting missions of COMECON countries

Year	Country	Participants	Collected germplasm
1973	Czechoslovakia	East Germany and Czechoslovakia	cereal crops
1976	Poland	East Germany and Poland	140 accessions of various agricultural crops
1977	Czechoslovakia	East Germany and Czechoslovakia	312 accessions of various crops
	USSR (Georgia)	East Germany and USSR	50 accessions of various crops
	Bulgaria	Czechoslovakia and Bulgaria	cereal crops
1978	Poland	East Germany and Poland	219 accessions of cereal and leguminous crops
	Czechoslovakia	Czechoslovakia and USSR	1500 accessions of local varieties of cereal and forage crops and wild species
1980	East Germany	East Germany and Poland	326 accessions of forage grasses and other crops
	Poland	Poland and USSR	74 accessions of local varieties of cereals and wild forage plants.
1981	Czechoslovakia	East Germany and Czechoslovakia	594 accessions of various agricultural crops
	USSR (Krasnodar and Stavropol Reg., North Ossetia, Daghestan, Azerbaijan and Georgia)	USSR and Poland	350 accessions of local varieties of cereal, leguminous and forage crops
1984	Poland	East Germany and Poland	78 accessions of various crops
1985	Mongolia	East Germany and Mongolia	20 accessions of various crops
1986	Bulgaria	Bulgaria and Hungary	233 accessions of cereal and fruit crops
	USSR (Georgia, Armenia and Azerbaijan)	USSR, Czechoslovakia and Poland	260 accessions of *Aegilops* spp., *Triticum* spp., *Hordeum* spp., leguminous, vegetable and forage crops
1987	Mongolia	East Germany and Mongolia	13 accessions of various crops
	Poland	Poland, Czechoslovakia and Bulgaria	150 accessions of grain legumes, forage plants and vegetables

Year	Country	Participants	Collected germplasm
1987	Czechoslovakia	Czechoslovakia, Poland, Bulgaria and Hungary	140 accessions of forage crops and wild relatives
1988	USSR (Uzbekistan, Tadjikistan, Kirghizia and Kazakhstan)	Czechoslovakia, Bulgaria, Poland, East Germany and USSR	105 accessions of vegetables, including onion (42) and garlic (18)
	Bulgaria	Czechoslovakia and Bulgaria	73 accessions, including 54 of wild *Triticum* spp. and *Aegilops* spp.
	East Germany	East Germany and Czechoslovakia	22 accessions of grasses
1989	Bulgaria	Bulgaria, USSR and Poland	125 accessions, including *Aegilops* spp.
	USSR (Azerbaijan and Armenia)	Czechoslovakia, Bulgaria, USSR and Poland	200 accessions of forage grasses and cereals (a lot of *Aegilops* spp. accessions)
	Poland	Czechoslovakia, Bulgaria and USSR	140 accessions of cereal, leguminous and oil crops
1990	Poland	Czechoslovakia, Bulgaria and USSR	57 accessions of leguminous and forage crops
	USSR (Altai Mt. and Western Siberia)	Czechoslovakia, Poland and USSR	112 accessions of *Allium* spp.
	USSR (Kazakhstan and Uzbekistan)	Czechoslovakia, Poland, Bulgaria and USSR	185 accessions of cultivated and wild forage crops and 300 herbarium specimens
	Mongolia	Czechoslovakia, Mongolia, Bulgaria and USSR	200 accessions of forage crops and cereals (wheat), and 237 herbarium specimens
	Poland	Poland and Czechoslovakia	36 accessions of forage grasses

Appendix VIII. International COMECON descriptor lists

1. The International COMECON List of Descriptors for the genus *Triticum* L. Leningrad, 1974, USSR.
2. The International COMECON List of Descriptors for the genus *Lycopersicon.* Tourn. Leningrad, 1979, USSR.
3. The International COMECON List of Descriptors for the species *Solanum melongena* L. (genus *Solanum* (Tourn.) L.). Leningrad, 1979, USSR.
4. The International COMECON List of Descriptors for the species *Allium cepa* L. Olomouc, 1980, CzSSR.
5. The International COMECON List of Descriptors for the species *Cucumis sativus* L. Leningrad, 1980, USSR.
6. The International COMECON List of Descriptors for the species *Cucurbita pepo* L. var. *giraumontia* Duch. Leningrad, 1980, USSR.
7. The International COMECON List of Descriptors for the genus *Pisum* L. Leningrad, 1981, USSR.
8. The International COMECON List of Descriptors for the genus *Faba* Mill. Leningrad, 1981, USSR.
9. The International COMECON List of Descriptors for the species *Brassica rapa* L. and *Brassica napus* subsp. *rapifera* Metzg. Leningrad, 1982, USSR.
10. The International COMECON List of Descriptors for the genus *Beta* L. Leningrad, 1982, USSR.
11. The International COMECON List of Descriptors for the genus *Sorghum* Moench. Leningrad, 1982, USSR.
12. The International COMECON List of Descriptors for the species *Panicum miliaceum* L. Leningrad, 1982, USSR.
13. The International COMECON List of Descriptors for the species *Vicia sativa* L. Leningrad, 1983, USSR.
14. The International COMECON List of Descriptors for the genus *Lupinus* L. Leningrad, 1983, USSR.
15. The International COMECON List of Descriptors for the genus *Trifolium* L. Leningrad, 1983, USSR.
16. The International COMECON List of Descriptors for the genus *Secale* L. Leningrad, 1984, USSR.
17. The International COMECON List of Descriptors for the genus *Hordeum* L. Leningrad, 1984, USSR.
18. The International COMECON List of Descriptors for the genus *Avena* L. Leningrad, 1984, USSR.
19. The International COMECON List of Descriptors for the species *Zea mays* L. Leningrad, 1984, USSR.
20. The International COMECON List of Descriptors of the potato species of the section *Tuberarium* (Dun.) Buk. of the genus *Solanum* L. Leningrad, 1984, USSR.
21. The International COMECON List of Descriptors for the genus *Medicago* L. subgen. *medicago* – subgen *Falcago* (Reichb.) Peterm. Leningrad, 1984, USSR.
22. The International COMECON List of Descriptors for the genus *Phaseolus* L. Leningrad, 1984, USSR.

23. The International COMECON List of Descriptors for the genus *Lens* Mill. Leningrad, 1984, USSR.
24. The International COMECON List of Descriptors for the family *Poaceae* Barnh. (of the genera: *Phleum* L., *Festuca* L., *Dactylis* L., *Lolium* L. and other genera of perenial grasses). Leningrad, 1985, USSR.
25. The International COMECON List of Descriptors for the species *Brassica oleraceae* L. var. capitata L. Leningrad, 1986, USSR.
26. The International COMECON List of Descriptors for the species *Capsicum annum* L. Leningrad, 1986, USSR.
27. The International COMECON List of Descriptors for the genus *Heliantus* L. Leningrad, 1988, USSR.
28. The International COMECON List of Descriptors for the genus *Persica* Mill. Leningrad, 1988, USSR.
29. The International COMECON List of Descriptors for the genus *Prunus* L. Leningrad, 1988, USSR.
30. The International COMECON List of Descriptors for the subfamily *Maloideae* (Malus Mill., Pyrus L., Cydonia Mill.) Leningrad, 1989, USSR.
31. The International COMECON List of Descriptors for the species *Cannabis sativa* L. Leningrad, 1989, USSR.
32. The International COMECON List of Descriptors for the species *Linum usitatissimum* L. Leningrad, 1989, USSR.
33. The International COMECON List of Descriptors for the genus *Citrullus* Schrad. Leningrad, 1989, USSR.
34. The International COMECON List of Descriptors for the species *Cucumis melo* L. Leningrad, 1989, USSR.
35. The International COMECON List of Descriptors for the genus *Cucurbita* L. Leningrad, 1989, USSR.
36. The International COMECON List of Descriptors for the genus *Gossypium* L. Leningrad, 1990, USSR.
37. The International COMECON List of Descriptors for the genus *Armeniaca* Hop. Leningrad, 1990, USSR.
38. The International COMECON List of Descriptors for the genus *Cerasus* Mill. Leningrad, 1990, USSR.
39. The International COMECON List of Descriptors for the genus *Glycine* Willd. Leningrad, 1990, USSR.
40. The International COMECON List of Descriptors for the species *Vitis vinifera* L. Leningrad, 1991, USSR.
41. The International COMECON List of Descriptors for the genus *Ribes* L. subgenus *Ribesia* (Berl.) Jancz. Leningrad, 1993, USSR.
42. The International COMECON List of Descriptors for the genus *Grossularia* (Tourn.) Mill. Leningrad, 1993, USSR.

Appendix IX. Establishment of genebanks in COMECON countries

The Eastern European countries, which had been the theatre of combat during World War II, faced the problem of restoration and development of their national economies which involved a need for self-supporting agricultural production. An important part in the solution of this problem was played by the collections of genetic resources held by each country. However, the plant genetic resource collections which existed before the war were either lost, or had suffered a large reduction in germination ability. N.I. Vavilov's ideas, his theoretical concepts, and the activities of VIR in the 1950s, 1960s and 1970s played an important role in the organization and development of genebanks in Eastern European countries, as did FAO, UNEP and IBPGR.

Bulgaria

The activities in the field of plant introduction in Bulgaria were initiated by Academician D. Kostov, who had worked in VIR for seven years and was one of the first followers and disciples of N.I. Vavilov. In 1940, he founded the Department of Ecology and Introduction at the Central Institute of Agriculture in Sofia, having launched the active study of plant diversity in the country. The Department functioned for some years but was closed because insufficient attention was paid at that time to the introduction and study of plant resources. The collected diversity of plants was irrevocably lost.

In 1952, the Department of Plant Introduction was organized at the Institute of Plant Industry, where the former research was restarted. The first data on the systematized preservation and documentation of seed collections date back to 1953. In 1956, the first seed-collecting missions aimed at local and primitive varieties were held in various parts of the country. This collection was the base for the creation, in 1977, of the Institute of Introduction and Plant Resources in Sadovo, which received the status of a national centre, responsible for organization and execution of scientific activity in the field of collecting, study and preservation of foreign and local plant resources for the purpose of use in plant breeding programmes of the country.

Major objectives of the Institute are:

- collecting, study, evaluation and storage of wild forms, populations and local varieties of the country's crop plants
- preservation of plant breeding materials with valuable biological and economic characters
- evaluation of the ecological conditions of Bulgaria for cultivating the world's best improved varieties and their direct introduction into national agriculture.

The collections have been mainly enriched through international exchange and expeditions within the country, as well as by adding cultivars and lines bred in Bulgaria. Since the mid-1980s, experts at the Institute have conducted joint expeditions with representatives of national institutes in the territories of COMECON countries. As Bulgaria is located in the geographical region favourable for almost all agricultural crops of a moderate climate, the Institute has established new fruitful connections with many countries of the world for plant genetic resource exchange. The Institute carries out domestic expeditions in compliance with a collecting programme, attributing great importance to landraces, populations and local varieties still existing in the country.

As a result, over 2200 local samples of various crops have been collected: corn, durum wheat, beans, melon, pepper and watermelon. In mountain areas, various wild species of alfalfa, *Lotus, Dactylis glomerata, Poa pratensis,* white and red clover occur. Some species of *Elymus, Aegilops,* etc. can be found in the Black Sea coastal regions. Some landraces of different crops may be used either directly in agricultural production, or after improvement. The genetic diversity from the collection of the Institute is widely used in agriculture. Bulgarian cultivars of the main field crops, i.e. wheat, sugar beet, tobacco, cotton and beans, occupy more than 95% of the sowing area in Bulgaria. The same situation exists with cultivars of tomato, pepper and grapes. The results of the participation of the country in international variety testing show a high level of breeding of various agricultural crops, and the contribution of the national programme to plant genetic resources is far from being the least of these (Alexanian 1997). Since 1982, the Institute collection has been conserved at 18°C in the National Genebank. The new National Seed Store with a capacity of 400 000 samples was opened in 1984. At present, the Bulgarian Collections consist of about 40 000 samples.

Hungary

From the beginning of the 20th century, landraces, ecotypes and other locally adapted plant materials have been involved in scientific breeding conducted in Hungary. After World War II, it was essential to restore the breeders' working collections along with rebuilding the breeding stations, both of which were almost completely destroyed during the war. The Hungarian genebank was started in 1948/1949 with the collection of valuable landraces of maize, red clover and alfalfa. In 1952, Professor A. Janossy organized the National Institute for Variety Testing and was appointed its first director. He was the founder of the Hungarian collection of plant genetic resources based on the theories of N.I. Vavilov.

In the 1950s, owing to the increasing volume of breeding, large amounts of plant material were accumulated in the collections, though not used in breeding programmes, despite the fact that they possessed some interesting characters useful for future breeding. At that time the Section for Evaluation and Maintenance of Raw Materials Collections of grain crops and legumes was first established by the Agricultural Institute at Martonvasar and by the Agricultural School at Szentes.

In 1954, the Central Department of Variety Collection Maintenance was opened at Tapioszele as an independent part of the Institute for Variety Testing. This Department continued to perform exploration and collection of old improved varieties and other plant materials under the leadership of A. Janossy. The greater part of the collection was represented by Hungarian landraces, populations and wild relatives of cultivated plants collected by Janossy.

The old wheats collected in different regions of Hungary were evaluated for the first time in 1955. A considerable part of the valuable maize ecotypes of the Hungarian basin were collected and saved in the 1950s before the rapid spread of hybrids. In the late 1950s, exploration and collecting of local bean varieties were initiated. In 1957 all collections comprised of different crops were moved to the Department at Tapiószele, thus launching centralized work on plant resources in Hungary.

As a consequence of the growth of plant resources collections representing basic breeding materials, the Department at Tapiószele was, in 1959, reorganized into the National Institute of Agrobotany under the directorship of Janossy. The major task of the Institute was to improve and maintain the plant germplasm collections in order to substantiate research on agrobotanical problems associated with the improvement of plant production in Hungary (Janossy 1974).

In 1976 the Institute of Agrobotany was affiliated to the National Institute for Agricultural Variety Testing. By a decision of the Ministry of Agriculture and Food, this Institute acquired the status of the national collection of plant genetic resources. From 1993, the Institute of Agrobotany, as an independent institution, became the coordinator of the activities related to plant genetic resources within the country, and the national genebank.

The following tasks were specified as the main priorities for the genebank: introduction of genetic variability of cultivated plants and their wild relatives; regeneration and long-term storage of the most valuable plant cultivars; evaluation of the initial breeding materials; creation of a modern database for the stored germplasm. The main purpose of the Institute of Agrobotany is to preserve the national plant genetic diversity and attract valuable foreign plant breeding materials. To implement maintenance and preservation of plant genetic resources, a long-term storage facility at a temperature of +4°C was established in 1973 at Tapiószele. The plan, based on the annual seed collecting missions and the explorations of the country's territory, envisages adding to the collection no less than 2000 samples of various agricultural species, which are still absent from the genebank.

In 1993, the Institute's collection included 54 500 accessions, over 16 000 of which were of Hungarian origin. The base collection at (–20°C) consisted of more than 7000 accessions, with 39 500 accessions maintained in the active collection at (0°C).

The national programme on plant genetic resources also involves a number of other institutes, the activities of which are coordinated by the Institute of Agrobotany. These institutes are located in various regions of the country and maintain the collections of certain crops. The national network includes the Horticultural Institute in Budapest (5430 samples of fruit-bearing plants), the Institute of Viticulture and Viniculture in Budapest (2100 samples of grapes), and the Cereal Institute in Szeged (2800 samples of wheat).

Since 1984, the main role in forming the national plant genetic resources collection has been played through the cooperation of the COMECON countries. The crop curators attended meetings on plant descriptors where the lists of plant characters and other aspects of evaluation methodologies were extensively discussed and standardized by the participating countries.

East Germany

Inspired by the ideas of N.I. Vavilov and especially by the lecture on genetic centres which he gave at the International Genetics Congress in Berlin in 1927, German researchers, headed by E. Baur, began conceptual work towards establishing an institute for plant genetic resources. The genebank has been named after N.I. Vavilov since its establishment.

After long and difficult negotiations, but with strong support from F. von Wettstein, a senator from the Kaiser Wilhelm Society (now the Max Planck Society), a new institute was finally founded in Tuttenhof near Vienna in 1943. The first director of the Institute of Cultivated Plant Research was H. Stubbe (1902-1989), who soon had to transfer the new institute to the Harz Mountains in central Germany because of the outbreak of the Second World War. After the war the new location turned out to be in the Soviet Zone of occupation, and the Soviet military administration offered the institute a large estate at Gatersleben, in a traditional seed-growing area close to Quedlinburg, on the northern rim of the Harz Mountains. The Institute has been in this area since 1946.

From the beginning, germplasm collections played an important role in the Institute's programme, starting with about 3500 accessions in 1945/46. These initial collections were part of the Systematics and Collections Department. R. Mansfeld (1901-1960), who had been the head of this Department since 1949, began the work on maintaining large collections of cultivated plants and their wild relatives. In 1969 the Genebank Department, headed by Chr.O. Lehmann (1926-1992), was administratively separated from the systematics group.

The state collection of the former GDR from after the Second World War and up to the reunification of Germany in 1990 was preserved in the Central Institute of Genetics and Crop Plant Research in Gatersleben, which had received the status of national curator. The study of the material was performed in cooperation with the special institutes, universities and plant breeding centres. The potato collection was preserved in the Institute of Potato-growing in Gross Lüsewitz (over 4000 samples), and fruit crops in the Institute of Fruit Crops in Dresden-Pillnitz (about 1000 samples), both of which have been subsidiaries of the Gatersleben genebank since 1992.

Collecting missions played an important role in the development of the Gatersleben genebank. Up to the present, the genebank has been maintaining potato samples collected by E. Bauer and R. Schick in South America in 1930, and accessions of different crops collected in Turkey, India, Nepal, Ethiopia, Albania and Greece before the Institute was founded. Immediately after the Second World War, large collections formed by German scientists in the pre-war period were added to the genebank. Despite financial and organizational problems, collecting missions started in 1950. Total collections made in South America by Dr Hans Ross, were held at Braunschweig. Subsamples were also provided to Gross Lüsewitz.

Germplasm activities were activated in 1974, with the influence of FAO and the Eucarpia Genebank Committee. Specific attention was given to collecting wild species and landraces. A distinctive feature of these explorations was the fact that the country signed intergovernmental agreements for a three-year period with the national genetic resources centres of each country surveyed. After that, the collecting was performed along with careful study of the areas involved in the search for particular crops. All the materials collected were divided into equal parts. One half remained in the explored country. On this basis, joint seed collecting missions were undertaken in Italy (1980-1992), North Korea (1985-1992), Austria (4 missions), Cuba (1986-1993), Libya, China (up to 3 expeditions). At that time, 10 collecting missions took place in Georgia, which enriched the collection with more than 2500 accessions of small grains, vegetables, leguminous crops, etc. (Hammer 1993).

Nearly 10 000 samples were collected within the period 1974 to 1990. The collections started in Central Europe and gradually extended to the Mediterranean, Central African territories, Eastern Asia and Latin America. After the reunification of Germany the genebank was re-established as the Institute of Plant Genetics and Crop Plant Research. Now the genebank at Gatersleben, with three refrigerating chambers with storage temperatures of +4°C, 0°C and –15°C, houses a collection exceeding 10 000 accessions. *In vitro* conservation at low temperatures and cryopreservation are also being developed.

Poland

Collecting and conservation of plant genetic resources were established in Poland by Professor L. Kasnowski in the Research Institute of Agronomy (PINGW) at Pulawy in 1922 and the Agricultural Academy at Dublany. From the 19th century, the practical accomplishments of Polish breeding were of a high level. The theoretical basis of plant breeding was established by K. Miczynski, A. Sempolowski, E. Zaleski and E. Malinowski. In 1923, 84 Polish seed firms existed and 345 varieties of cultivated plants were registered in the "Catalogue of Original Varieties". Some of the companies used Kasnowski's material and some had their own collections of initial breeding materials, based on landraces and old populations intensively used in their plant breeding programmes.

During World War II the collections were partly lost, and work in the field of collecting and preservation of plant genetic resources was restarted only in 1951, when the Plant Breeding and Acclimatization Institute (IHAR) was founded. The Institute has continued the tradition of PINGW and preserved its collection. The main tasks of the Institute was research on the valuable characters of the collections.

In the mid-1970s, Polish scientists became aware of the increasing genetic erosion and disappearance of the rich diversity of cultivated plants in the country. Since 1971, several independent collecting missions have been sent to different parts of Poland to collect cultivated, wild and weedy plant germplasm. In order to preserve the diversity still existing, a special scientific body, the National Department of Plant Genetic Resources, was established in 1979 at the Plant Breeding and Acclimatization Institute. This Institute, being the national plant genetic resources coordinator, has been responsible for scientific management of Polish plant genetic resources collections, preserved in separate institutes.

The main tasks have been as follows: collecting and primary evaluation of various genotypes of agricultural plants and their long-term storage in viable conditions in the national genebanks; supplying the country's breeders with initial plant breeding materials and exchange of seed samples with foreign institutions; creation of a database on the genebank and providing germplasm information to all interested users; organization of seed collecting missions in Poland and in other countries (Podyma 1993).

In 1981, the first National Programme on collecting and preservation of plant genetic diversity of Poland was adopted, the scientific management of which was vested in the Department. Due to their active efforts, plant genetic resources experts managed in a short time to collect a large quantity of plant samples representing various landraces of agricultural crops and their wild relatives. Up to 1991, the Institute organized more than five international seed collecting missions, which resulted in the addition of 1300 accessions to the national collection of Poland.

Within the framework of the Programme, a study has taken place on material belonging to 57 genera, concentrated in more than 10 Institutes. The greatest collection is studied and conserved at IHAR (all kinds of cereals, leguminous, industrial and fodder crops). A large collection of rye, numbering about 1300 wild forms and about 900 cultivar populations, has been maintained at the Botanical Garden of the Polish Academy of Sciences.

Germplasm exchange with foreign countries has also contributed to increasing the collection. Collecting missions seldom travelled abroad, but when they did they included a scientist from IHAR in the exploration teams staffed with representatives of national universities. The territories of the COMECON countries were the most thoroughly explored.

Since 1987, the IHAR national genebank collections have been preserved in the long-term storage facilities consisting of two chambers at –15°C and four at +4°C. About 40 000 accessions are conserved in the long-term storage facilities, while the total number of accessions of various agricultural crops and their wild relatives in the national collection is 60 000. In recent years, Poland has actively participated in international programmes on plant genetic resources and maintained close links with international centres and organizations.

Czech Republic and Slovakia

Studying, conservation and utilization of plant genetic resources has been a long-standing tradition in Bohemia and Moravia. Various research and breeding stations as well as botanical gardens were already working with genetic resources at the beginning of this century. Collections of a wide range of plant species were concentrated in three places as follows:

1. The Husbandry Botanic Station in Tabor started testing genetic resources of barley in 1899, and wheat and other crops in 1903.
2. The Chemical and Physiological Station in Jenec near Prague, established in 1898, tested different crops and varieties. In 1920 the Station was transferred to Uhrineves at the National Institute of Agriculture, Prague. This Institute took over the collection of the Husbandry Botanic Station in Tabor. In 1948 it was transferred to Doksany and in 1952 to the Institute of Crop Production in Praha-Ruzyně.
3. The Moravian Land Agricultural Institute in Brno, established in 1919, contains a collection of landraces. From 1951 to 1954, this collection was transferred to newly established institutions. A greater part went to the Institute of Cereals in Kromeriz, to the Institute of Fodder Crops in Troubsko and to the Grassland Station in Roznov.

Like the Czech institutes, the Slovakian National Agricultural Institutes in Bratislava and Kosice were carrying out studies on genetic resources after the First World War. Foreign varieties were introduced to Slovakia in 1900, but documentation is missing. Together with landraces, local varieties were used in breeding. Landraces and improved varieties were also tested in several great estates in Sladkovicovo and breeding stations. In Sladkovicovo, for instance, 200 wheat varieties with good resistance to rust and mildew were accumulated by 1943. Resources for breeding were assembled in these significant working sites of Slovakia. After 1951, a greater part of them was transferred to the collections of newly established institutes.

At that time, the collections of the Research Institute of Crop Production in Prague-Ruzyne contained 2847 samples of cereals, legumes, oil and fodder crops. A number of landraces and breeding varieties of Czechoslovak origin from the beginning of the 20th century have been preserved in the collections of the above-mentioned institutes. Old varieties of grapes and fruit-trees from the same period have been preserved in plantations registered or documented by these institutes. In the fifties, genetic resources of cultivated plants were taken over by today's research institutions established by that time. The original amount of 6000 varieties stored in 1951 in the Czechoslovak collections was growing rapidly, especially in the 1950s and 1960s.

Thus, it may be assumed that the study and exploration of the country's genetic diversity started in 1951, when a government decision established a system of institutions dealing with plant science and breeding. Since 1954, an appreciable role in this sphere has been played by the Institute of Crop Production (Praha-Ruzyně) in the Czech Republic, and by the Institute for Plant Production (Piestany) in Slovakia, both of which have been subsequently nominated as the national coordinators of plant genetic resources, while the first has been preserving the Czechoslovak collection.

The national system of plant genetic resources of these countries included 28 research institutions studying separate crops or groups of crops, and a number of plant breeding stations. The largest Institutes in the national system are: the Plant Breeding Institute of Vegetables in Olomouc (about 10 000 samples of vegetables and aromatic plants); the Institute of Cereals in Kromeriz (over 10 000 samples of wheat, barley, rye and oats); the Institute of Industrial and Pulse Crops in Sumperk (more than 4000 samples of pea, vetch, bean, flax, etc.); and the Institute for Plant Production in Piestany (now the Slovak Republic) (over 4500 samples of wheat, barley, triticale, beans, alfalfa, clover, etc.).

A new stage in plant genetic resources activities in Czechoslovakia was its participation since 1964 in the COMECON country cooperation, which began to develop from 1973, when the process of international coordination of the collections preserving cultivated and wild plant species was launched. Coordination of activities in the sphere of plant genetic resources has been carried out by the Czechoslovak Council on Genetic Resources of cultivated plants at the Institute of Crop Production. The purpose of the Council is to enhance study, preservation and utilization of the genetic diversity, and coordination with international and foreign national programmes.

The specific role in collecting, studying and preservation of plant genetic diversity has been played by the Department of Plant Genetic Resources at the Institute of Crop Production. Almost 80% of plant diversity which reaches establishments in the country and abroad, is sent by the Department. Conserved genetic diversity increased from 6000 samples in 1951 to 43 000 samples in 1987, 7.2% of which are Czechoslovak landraces and improved varieties (Dotlacil, 1993).

The collections have been enriched by germplasm collecting missions that explored its own territory, exchange of seed samples of germplasm and improved varieties with foreign countries, and development of new Czechoslovak breeding lines. Participation in joint collecting missions organized by COMECON in the member countries provided a good source of additions to the collections.

In 1989, scientific accomplishments in the field of long-term storage enriched by international experience allowed the country to construct the National Genebank where 10 000 accessions were placed for long-term storage. Research in the sphere of cryopreservation has been actively conducted. This will make it possible to store the accessions for a 100 years or more at a lower cost. The national plant genetic diversity is stored in the genebank of the Institute of Crop Production. The specialized Institutes of national networks of plant resources have special chambers for storage of the working collections and breeding material.

Romania

The programme of crop genetic diversity conservation was started in Romania in 1975 with organization of a group responsible for collecting and preserving local maize varieties at the Institute of Cereal and Industrial Crops, at Fundulea. Romania cooperated with COMECON countries in the field of plant genetic resources. In 1986 the country officially left the programme of joint research. Nevertheless, it is recognized that the establishment of the national collection was the result of such cooperation. In 1990, a plant Genebank was established in Suceava and the management centre of the Romanian National Committee on Plant Genetic Resources was located there. Its objectives are: collecting, evaluation and conservation of plant genetic resources. The collecting priority is Rumanian landraces, wild species and, as far as possible, the relatives of cutivated plants. The necessary conditions exist for medium-term storage (4°C) and long-term storage (below –20°C) in very modern refrigeration installations. *In vitro* medium-term storage of potato is also carried out (Frison *et al.* 1995). At present, the country's germplasm collection has reached 10 450 accessions of various agricultural crops, the major part of which is represented by *Sorghum* (3600 accessions), barley (2280), and leguminous crops (1750). Duplicate collections are placed at the Department of Plant Genetic Resources of the Institute of Cereal and Industrial Crops, Fundulea, and at the genebanks of Russia and Germany (Alexanian 1997).

Yugoslavia

In the former Yugoslavia, there was a wide range of maize varietal populations well adapted to the local environments. Although plant breeding has a long tradition in

Yugoslavia, the organization of work on plant genetic resources is recent. In the 1960s, the Institute of Maize "Zemun Pole" together with other plant breeding institutes began collecting, studying and preserving local maize diversity. By 1969, over 2000 samples were collected in all the regions of Yugoslavia. Since 1986, the Institute has received the status of a genebank and became the national coordinator of the plant genetic resources programme. A total of 34 000 accessions were processed and databases were established for the needs of the Genebank. All seed samples are temporarily stored in the chambers of the Institute at 4°C. Accessions of fruit trees are maintained under *in situ* and *ex situ* conditions controlled by a specialized institute.

Yugoslavia has created a Council of Experts for the coordination of activities and development of the national programme, as well as a Committee on Agriculture situated in the Federal Government with the purpose of development and financing of the programme on plant genetic resources.

The main institutes, from which the national programme has been developed are the Maize Research Institute "Zemum Polje" (maize, sorghum and millet), the Vegetable Research Centre in Smederevska Palanka, the Institute of Agriculture in Lyublyana (potato), the Institute of Tobacco in Prilepa and the Fruit Research Centre in Cacak. The national network has also incorporated departments of a number of agricultural colleges. The Maize collection is the largest in Yugoslavia; 12 000 maize accessions are stored in the Institute of Field Crops Breeding at the Faculty of Agronomy, Zagreb University.

However, interregional conflicts and wide-scale military activities on the territory of the country with no definite prospects of ceasing have deferred the implementation of these proposals (Alexanian 1997).

Mongolia

The programme on plant genetic resources in Mongolia dates back to the collecting mission undertaken in the country in 1922 by Pisarev at Vavilov's initiative. All samples collected in Mongolia were preserved in the VIR collection. From the start of cooperation within the COMECON network, Mongolia commenced the work of collecting, preserving and studying plant genetic resources, which had never been performed before, despite Mongolias representing a centre of diversity of cereal and forage crops. The Institute of Plant Industry and Crop Cultivation in Darhan acquired the status of the national genebank, and Mongolian experts started making collections of the crops most important for Mongolian agriculture in close cooperation with their colleagues from the USSR. Mongolia was an active participant in the work performed within the COMECON STC network, being involved in all fields of research.

Bibliography

Alexanyan, S.M. 1992. N.I. Vavilov Institute of Plant Industry: problems of mobilization and conservation of plant genetic resources. Country report. Rural development course, July 1-11, Seul, Korea, p. 203-210. [English].

Alexanyan, S.M. 1994. Prospects of development of *ex situ* conservation of plant genetic resources collections in Russia. Pp. 70-75 *in* Integration of conservation strategies of plant genetic resources in Europe. Proceeding of International Symp. on PGR in Europe, Gatersleben, Germany, 1993. [English].

Alexanyan, S.M. 1998. The modern problems of mobilization and preservation *ex situ* of bio-diversity of cultivated plants. PhD thesis. N.I. Vavilov Institute of Plant Industry. St. Petersburg. [Russian].

Alexanyan, S.M. and V.P. Denisov. 1988. World scientists join in Soviet celebration of N.I. Vavilov centennial. Diversity 15:5. [English].

Alexanyan, S.M. and G.G. Heintz. 1989. Council for Mutual Economic Aid Commission establishes Scientific Technical Council for Plant Genetic Resources. Diversity 5(2-3):9. [English].

Bakhareva, S.N. 1988. Rastitelnye resursy Zapadnoi i Tzentralnoi Afriki. [Plant resources of Western and Central Africa]. L., p. 236. [Russian].

Bakhareva, S.N. 1990. The USSR policy for exchanging genetic resources and the germplasm collection procedures of the Vavilov Institute. Diversity 6(3-4): 10. [English].

Bakhteev, F.Kh. 1988. Nikolay Ivanovich Vavilov. 1887-1943. Nauka Press, Novosibirsk, 270 p. [Russian].

Banyai, L. 1977. Andor Janossy (1908-1975). Agrobotanika 17:13- 25. [Hungarian, English].

Bares, I. and J. Sehnalova. 1981. Preservation of landraces of cultivated plants in Czechoslovakia. Kulturpflanze XXIX:67-77. [English].

Bateson, W. 1925. Science in Russia. Nature, November:1-8. [English].

Bazilevskaya, N.A. 1964. Teoriya i metody introduktzii rastenii. [The theory and methods of introduction of plants]. Moscow, 131 p. [Russian].

Biokhimiya kulturnykh rastenii. 1936. Khlebnye zlaki. [Biochemistry of cultivated plants. Cereals]. 1st edn. vol. 1., M.-L., 315 p. [Russian].

Biokhimiya kulturnykh rastenii. 1938. Maslichnye kultury. [Biochemistry of cultivated plants. Oil crops]. vol. 3, M.-L., 398 p. [Russian].

Biokhimiya kulturnykh rastenii. 1938. Ovoshnye i bakhchevye [Biochemistry of cultivated plants. Vegetable crops]. vol. 4, M.-L., 450 p. [Russian].

Biokhimiya kulturnykh rastenii. 1938. Tekhnicheskie kultury. [Biochemistry of cultivated plants. Fiber crops]. vol. 5, M.-L., 288 p. [Russian].

Biokhimiya kulturnykh rastenii. 1938. Zernobobovye i kormovye kultury. [Biochemistry of cultivated plants. Legumes and forages). 1st edn. vol. 2, M.-L., 420 p. [Russian].

Biokhimiya kulturnykh rastenii. 1940. Plodobye i yagodnye. [Biochemistry of cultivated plants. Fruit and berry crops]. vol. 7, M.-L., 561 p. [Russian].

Biokhimiya kulturnykh rastenii. 1948. Problemy rastitelnykh veshestv. [Biochemistry of cultivated plants. The problems of plant substances]. vol. 8, M.-L., 710 p. [Russian].

Biokhimiya kulturnykh rastenii. 1958. Khlebnye i krupyanye kultury. [Biochemistry of cultivated plants. Cereals]. 2nd edn. vol. 1, M.-L., 701 p. [Russian].

Biokhimiya kulturnykh rastenii. 1961. Ovoshnye kultury. [Biochemistry of cultivated plants. Vegetable crops]. 2nd edn. vol. 2 M.-L., 544 p. [Russian].

Boos, G.V., G.V. Badina and V.I. Burenin. 1990. Geteroziz ovoshnykh kultyr. [Heterosis of vegetable crops]. Leningrad, 223 p. [Russian].

Brat'ya Nikolai i Sergei Vavilovy [The brothers Nikolai and Sergei Vavilovs]. 1994. Moscow, 46 p. [Russian].

Budin, K.Z. 1986. Geneticheskie osnovy selektzii kartofelya. [Genetical bases of potato breeding]. Leningrad, 192 p. [Russian].

Budin, K.Z. 1992. The USSR potato collection: its genetic potential and value for plant breeding. Diversity 8(1):12. [English].

Bukasov, S.M. 1928. Kulturnye rasteniya Meksiki, Gvatemaly i Kolumbii [The cultivated plants of Mexico, Guatemala and Colombia]. [Works of Applied Botany, of Genetics and Plant Breeding] Supplement no. 47, 553 p. [Russian, English].

Bulinska-Radomska, Z. and S. Goral. 1991. Current status of genetic resources activities in Poland. Genetica Polonica 32(1-2):27- 35. [English].

Coordination Centre. 1983. Voprosy selektzii i genetiki zernovykh kultur [The problems of plant breeding and genetic of cereal crops]. facs. 1, Moscow, p. 475. [Russian].

Coordination Centre. 1985. Voprosy selektzii i genetiki zernovykh kultur [The problems of plant breeding and genetic of cereal crops]. facs. 2, Sofia, p. 278. [Russian].

Coordination Centre. 1987. Voprosy selektzii i genetiki zernovykh kultur [The problems of plant breeding and genetic of cereal crops]. facs. 3, Prague, p. 545. [Russian].

Coordination Centre. 1990. Voprosy selektzii i genetiki zernovykh kultur [The problems of plant breeding and genetic of cereal crops]. facs. 4, Berlin, p. 329. [Russian].

Czembor, H.J. 1989. Plant Breeding and Acclimatization Institute – coordinator of plant genetic resources programme. Polish gene bank Newsletter No. 1. [English].

Davenport, Ch.B. 1992. Letter. Not only national suicide but a blow in the face for civilization. Herald of the Russian Academy of Science 62(7):513-514. [English].

Dorofeev, V.F. 1972. Pshenitzy Zakavkazya. [The wheats of Trans- Caucases]. Works of Appied Botany, Genetics and Plant Breeding 47(1):5-202. [Russian].

Dorofeev, V.F. 1976. Pshenitzy mira. [Wheats of the world]. Kolos, L., 487 p. [Russian].

Dorofeev, V.F. 1986. Teoriya tzentrov proiskhozhdeniya kulturnykh rastenii N.I. Vavilova i ee razvitie. [N.I.Vavilov's theory of center of origin cultivated plants and its development]. Vestnik sel'skokhozyaistvennoi nauki 1:55-66. [Russian].

Dorofeev, V.F. 1987. Pshenitzy mira. [Wheats of the world]. edn. 2, Leningrad, 560 p. [Russian].

Dorofeev, V.F., G.G Geints and S.M. Alexanyan. 1987. Zadachi i nekotorye rezultaty sotrudnichestva po probleme sbora, izuchenia, sokhraneniya i ispolzovaniya v selektzii geneticheskikh resursov kulturnykh rastenii. [Goals and some results of cooperation on the problem of exploration, studies, conservation and utilization of cultivated plant genetic resources]. Plant genetic resources – their evaluation and using in plant breeding. Sadovo, p. 10-17. [Russian, English].

Dotlacil, L., J. Rychtarik and I. Bares. 1991. Plant genetic resources in CSFR. Plant Genetic Resources. Czechoslovak board on plant genetic resources. Annual report, p. 42-52. [English, Czech].

Dotlacil, L., Z. Stehno and M. Resatko. 1993. National activities on plant genetic resources in the Czech republic. Proc. Int. Symp. on Plant Genetic Resources Activities in Europe, Germany, 6-8 December. (In press). [English].

Dragavtsev, V.A., G.B Udovenko, N.F. Batygin, *et al.* 1995. Teoreticheskie osnovy selektzii rastenii. Fiziologichrskie aspekty prikladnoi botaniki, genetiki i selektzii. [Theoretical bases of plant breeding. Physiological aspects of applied botany, genetics and plant breeding]. Vol. 2, St-Petersburg, 622 p. [Russian, English].

Dragavtsev, V.A. and S.M. Alexanyan. 1991. The VIR network: problems of mobilization and conservation of plant genetic resources; the concept of international collaboration. Pp. 67-71 *in* Crop Networks. Paper of the EUCARPIA/IBPGR symposium held at Wageningen, The Netherlands, 3-6 December 1990. [English].

Dragavtsev, V.A. and S.M. Alexanyan. 1993. Dramatic changes in former Soviet Union require new approaches at VIR. Diversity 9(1/2). [English].

Enken, V.B. 1959. Soya. Moscow, 662 p. [Russian].

Flyaksberger, K.A. 1915a. Opredelitel pshenitz. [Determination of wheats]. Bull. of the Bureau of Applied Botany 8(1/2):3-190. [Russian].

Flyaksberger, K.A. 1915b. Obzor raznovidnostei pshenitzy Sibiri. [Survey of wheat's botanical varieties in Siberia]. Bull. of the Bureau of Applied Botany 8(8):557-862. [Russian].

Flyaksberger, K.A. 1922. R.E. Regel. Works of Applied Botany and Plant Breeding 12(1):3-24. [Russian].

Frankel, O. 1935. A letter to N.I. Vavilov of 2 July 1935. Personal Yu. N. Vavilov's archive. [English].

Frison, E.A. and J. Serwinski. 1995. Directory of European Institutions holding crop genetic resources collections. ECP/GR, FAO/IPGRI, Rome. 500 p. [English].

Gade, H. 1993. Beitrage zur Gerschichte der Pflanzenzuchtung und Saatgutwirtschaft in der funf neuen Bundeslandern Deutchlands. [German].

Geneticheskie resursy rastenii ikh izuchenie i ispolzovanie v selekztii. [The evaluation of plant genetic resources and their using in breeding]. 1987, Sadovo, Bulgaria, 116 p. [Russian, English].

Geneticheskie resursy rastenii ikh izuchenie i ispolzovanie v selekztii. [The evaluation of plant genetic resources and their using in breeding]. 1990, 178 p. [Russian, English].

Genetika kulturnykh rastenii. Pshenitza, yachmen, rozh. [Genetics of cultivated plants. Wheat, barley, rye]. 1986. Leningrad, 264 p. [Russian].

Genetika kulturnykh rastenii. Kukuruza, krupyanye, oves. [Genetics of cultivated plants. Maize, small grain, oats]. 1988. Leningrad, 276 p. [Russian].

Genetika kulturnykh rastenii. Zernobobovye, ovoshnye, bakhchevye. [Genetics of cultivated plants. Legumes, vegetables, melon crops]. 1990. Leningrad, 287 p. [Russian].

Goral, S., K. Hammer, P. Hanelt and C. Lehmann. 1985. Zum Gedenken an Wladyslaw Kulpa (1923-1984). Kulturphlanze 33:9-15, [German].

Gorbatenko, L.E. 1994. Teoriya introdukchii rastenii i ee voploshenie v deyatelnosti instituta. [Theory of plant introduction and its manifestation in the activity of the Institute]. Vestnik of the Russian Academy Agricultural Science 3:17-22. [Russian]

Grum-Grzhimailo, A.G. 1986. V poiskakh rastitelnykh resursov mira. Nekotorye nauchnye itogi puteshestvii akademika N.I.Vavilova. [Searching for plant resourses of the world. Some scientific results of academician N.I. Vavilov's trips]. Nauka Press, Leningrad, 152 p. [Russian].

Hall, A.D. 1930s. A reference of N.I. Vavilov. Archives John Innes Center. [English].

Hammer, K. 1993. The 50th anniversary of the Gatersleben genebank. Plant Genetic Resources Newsl. 91/92:1-8. [English].

Hammer, K. and F. Begemann. 1993. Gatersleben Genebank celebrates 50th anniversary. Diversity 9(1-2):22-23. [English].

Hammer, K., H. Gade and H. Knupffer. 1994. 50 Jahre Genbank Gatersleben – eine Ubersicht. Nutzbarmachung genetischer Ressourcen fur Zuchtung und Landschftsgestaltung. Vortrage für Pflanzenzuchtung, Heft 27:333-383. [German].

Hammer, K. and P. Hanelt. 1992. Christian O. Lehmann (1926-1992). Genetic Resources and Crop Evolution 39:61-69. [English].

Hanelt, P. 1994. Tradition und Fortschritt einer Forschungseinrichtung. Biol. Zent.bt 113:15-23. [German, English].

Hawkes, J.G. 1995. Academician N.I. Vavilov. Priroda [Nature] 12:63-67. [Russian].

Ivanov, A.I. 1980. Lyutzerna. [Alfalfa]. Moscow, Kolos, 350 p. [Russian].

Ivanov, A.I., Yu.D Soskov. 1983. Teoretizteskie osnovy introdukztii mnogoletnikh kormovykh rastenii. [Theoretical bases of the introduction of perennial forage plants]. Scientific bulletin of VIR, facs. 133, p. 13-20. [Russian].

Janossy, A. 1959. Foundation, task and activities of the National Institute for Agricultural Botany. Agrobotanika I:7-29. [Hungarian, Russian, German].

Janossy, A. 1974. Genebank work of the National Institute of Agrobotany. Agrobotanika 15:5-9. [Hungarian, English].

Kobylyanskaya, K.A. 1989. Poslevoennye zarubezhnye ekspeditzii VIR. [The VIR post-war plant collecting expeditions to foreign countries]. Works of Appied Botany, Genetics and Plant Breeding 126:19-27. [Russian]

Kobylyanskaya, K.A. 1990. Informatzia o vnutrisoyuznykh i zarubezhnykh ekspediztiyakh VIR v 1981-1985 gg. [Information of within-the-USSR and foreign expeditions VIR in 1981-1985]. N 562. L. 67 p. [Russian].

Kobylyansky, V.D. 1982. Rozh: Geneticheskie osnovy selekchii. [Rye: Genetic bases of breeding]. Kolos, Moscow, 271 p. [Russian].

Kobylyansky, V.D., A.G. Katerova and N.S. Lapikov. 1994. Development of initial material for breeding of hybrid rye in Russia. Russian J. Genetics 30(10):1403-1413. [English].

Konarev, V.G. 1994. Problems of species and genome in breeding. Russian J. Genetics 30(10):1293-1306. [English].

Konarev, V.G., I.P. Gavriluk, N.K. Gubareva, T.I. Peneva, *et al.* 1993. Teoreticheskie osnovy selektzii rastenii. Molekulyarno-biologicheskie aspekty prikladnoi botaniki, genetiki i selektzii. [Theoretical bases of plant breeding. Molecular biological aspects of applied botany, genetics and plant breeding]. Vol. 1, Kolos, M., 447 p. [Russian, English].

Korotkova, T.I. 1978. N.I.Vavilov in Saratov (1917-1921). Saratov., 120 p. [Russian].

Krivchenko, V.I. 1984. Ustoichivost zernovykh kolosovykh k vozbuditilyam golovnevykh boleznei. [The resistance of cereal crops to smut diseases]. Moscow, Kolos, 304 p. [Russian].

Krivchenko, V.I. 1988. The role of Vavilov in creating the national Soviet program for plant genetic resources. Diversity 16:5. [English].

Krivchenko, V.I. and S.M. Alexanyan. 1991. Vavilov Institute scientists heroically preserve world plant genetic resources collection during World War II siege of Leningrad. Diversity 7(4):10-13. [English].

Kulturnaya flora SSSR. 1935. Khlebnye zlaki. Pshenitza. [Cultivated flora of the USSR. Grain cereals. Wheat]. Vol. 1, M.-L., 435 p. [Russian].

Kulturnaya flora SSSR. 1936. Khlebnye zlaki. Rozh, Yachmen, Oves. [Cultivated flora of the USSR. Grain cereals. Rye, Barley, Oats]. Vol. 2, M.-L., 147 p. [Russian].

Kulturnaya flora SSSR. 1937. Zernovye bobovye. [Cultivated flora of the USSR. Grain legumes]. vol. 4, M.-L., 680 p. [Russian].

Kulturnaya flora SSSR. 1940. Orekhoplodnye kultury. [Cultivated flora of the USSR. Nut-Bearing Crops]. Vol. 17, M.-L., 354 p. [Russian].

Kulturnaya flora SSSR. 1940. Pryadilnye kultury. [Cultivated flora of the USSR. Fiber Crops]. Vol. 5, part 1, M.- L., 315 p. [Russian].

Kulturnaya flora SSSR. 1940. Yagodnye kultury. [Cultivated flora of the USSR. Small Fruit Crops]. Vol. 16, M.- L., 285 p. [Russian].

Kulturnaya flora SSSR. 1941. Maslichnye kultury.[Cultivated flora of the USSR. Oleiferous Plants]. Vol. 7, M.- L., 496 p. [Russian].

Kulturnaya flora SSSR. 1950. Mnogoletnie bobovye travy. Lutzerna, donnik, pazhitnik. [Cultivated flora of the USSR. Perennial Leguminous Grasses. Alfalfa, sweet clover, fenugreek]. Vol. 13, part 1, M.- L., 526 p. [Russian].

Kulturnaya flora SSSR. 1958. Ovoshnye paslenovye. (Tomat, baklazhan, chernyi paslen, dynnaya grusha, peretz, fizalis, mandragora). [Cultivated flora of the USSR. Vegetable Plants (Fam. Solanaceae – tomato, common eggplant, black nightshade, pepino pepper, husk - tomato, mandrake]. Vol. 20, M.-L., 531 p. [Russian].

Kulturnaya flora SSSR. 1971. Kartofel. [Cultivated flora of the USSR. Potato]. Vol. 9, Kolos, L., 448 p. [Russian].

Kulturnaya flora SSSR. 1971. Korneplodnye rasteniya. (Sem. Chenopodiaceae – svekla, sem.Umbelliferae – morkov, petrushka, selderei, pasternak). [Cultivated flora of the USSR.

Root Crops (Fam. Chenopodiaceae – beet, fam. Umbelliferae – carrot, parsley, celery, parsnip]. Vol. 19, Kolos, L., 436 p. [Russian].
Kulturnaya flora SSSR. 1975. Krupyanye kultury (grechikha, proso, ris). [Cultivated flora of the USSR. Groat Crops: buckwheat, millet, rice]. Vol. 3, Kolos, L., 364 p. [Russian].
Kulturnaya flora SSSR. 1978. Luk. [Cultivated flora of the USSR. *Allium*]. Vol. 10, Kolos, L., 264 p. [Russian].
Kulturnaya flora SSSR. 1979. Gorokh. [Cultivated flora of the USSR. Pea]. Vol. 4, Kolos, L., 324 p. [Russian].
Kulturnaya flora SSSR. 1979. Pshenitza. [Cultivated flora of the USSR. Wheat]. Vol. 1, Kolos, L., 346 p. [Russian].
Kulturnaya flora SSSR. 1982. Kukuruza. [Cultivated flora of the USSR. Maize]. Vol. 6, Kolos, M., 295 p. [Russian].
Kulturnaya flora SSSR. 1982. Tykvennye (Arbuz, tykva). [Cultivated flora of the USSR. Fam. Cucurbitaceae (Watermelon, pumpkin]. Vol. 21, part 1, Kolos, L., 279 p. [Russian].
Kulturnaya flora SSSR. 1983. Semechkovye (yablonya, grusha, aiva). [Cultivated flora of the USSR. Pome fruits (apple, pear, quince]. Vol. 14, Kolos, M., 320 p. [Russian].
Kulturnaya flora SSSR. 1984. Kapusta. [Cultivated flora of the USSR. Cabbage. Vol. 11, Kolos, L., 328 p. [Russian]
Kulturnaya flora SSSR. 1985. Korneplodnye rasteniya. [Sem. Brassicaeae – repa, turneps, bryukva, redka, redis). [Cultivated flora of the USSR. Root Crops (Fam. Brassicaeae – turnip, rutabaga, radish, small radish]. Vol. 18, Kolos, L., 324 p. [Russian].
Kulturnaya flora SSSR. 1988. Listovue ovoshnye rasteniya. (Sparzha, reven, shavel, shpinat, portulak, kress-salat, ukrop, tzikorii, salat). [Cultivated flora of the USSR. Leaves Vegetable Plants (Asparagus, rhubarb, sorrel, spinach, purslane, garden cress, dill, chicory, lettuce)]. Vol. 12, Kolos, L., 304 p. [Russian].
Kulturnaya flora SSSR. 1989. Rozh. [Cultivated flora of the USSR. Rye]. Vol. 2, part 1, Leningrad, 368 p. [Russian].
Kulturnaya flora SSSR. 1990. Yachmen. [Cultivated flora of the USSR. Barley]. Vol. 2, part 2, Leningrad, 421 p. [Russian].
Kulturnaya flora SSSR. 1993. Mnogoletnie bobovye travy. Klever, lyadvinetz. [Cultivated flora of the USSR. Perennial Leguminous Grasses. Clover, birds-foot trefoil]. Vol. 13, part. 2, Kolos, M., 335 p. [Russian].
Kulturnaya flora SSSR. 1994. Oves. [Cultivated flora of the USSR. Oat]. Vol. 2, part 3, Kolos, M., 367 p. [Russian].
Kulturnaya flora SSSR. Tykvennye (Oguretz, dynya). [Cultivated flora of the USSR. Fam. Cucurbitaceae (Cucumber, melon). vol. 21, part. 2, 1994, kolos, M., 288 p. [Russian].
Kuptsov, A.I. 1975. Vvedenie v geografiyu kul'turnykh rastenii. [Introduction to geography of cultivated plants]. Nauka Press, M., 295 p. [Russian]
Kuptsov, A.I. and M.E. Ramenskaya. 1989. Geograficheskie kontzeptzii N.I. Vavilova i sovremennost. [N.I. Vavilov's geographical concepts and modern time]. *In* N.I. Vavilov's Heritage in Modern Biology. Nauka Press, 147-155 p. [Russian]
Kurlovich, B.S., S.I. Rep'ev, *et al.* 1995. Teoreticheskie osnovy selektzii rastenii. Genofond i selektziya zernobobovykh kultur (lupin, vika, soya i boby). [Theoretical bases of plant breeding. The genebank and breeding of grain legumes (lupin, vetch, soya and bean]. Vol. 3, St-Petersburg, 438 p. [Russian, English].
Lemeshev, N.K. 1992. Meksikanskii ztentr proiskhozhdeniya i vidovogo raznoobraziya roda Gossypium i problema obogasheniya genofonda. [Mexican center of origin and Diversity of genus Gossypium and the problem development of genofund]. Doctoral dissertation. N.I. Vavilov Institute of Plant Industry, St-Petersburg. [Russian].
Levina, E.S. 1995. Vavilov, Lysenko, Timofeev-Resovsky... Moscow, 160 p. [Russian].
Levitskaya, N.G. and T.K. Lassan. 1992. Grigorii Andreevich Levitsky. Tzitologia [Cytology]

34(8):102-125. [Russian].
Litvinov, N.I. 1908. Pravila dlya proizvodstva odnoobraznykh posevov khlebnykh zlakov pri sravnitelno-botanicheskikh issledovaniyakh. [Guidelines for Production of Uniform Cereal Plantings during comparative Botanical Research]. Bull. of the Bureau of Applied Botany 1(1/2):86-89. [Russian].
Litvinov, N.I. 1915. O porazhenii yarovykh pshenitz zheltoi rzhavchinoi v Kamennoi Stepi v 1914 godu. [On the Puccinia glumarum incidence in spring wheats in Kamennaya Steppe in 1914]. Bull. of the Bureau of Applied Botany 8(6):808-815. [Russian].
Makashova, R.Kh. 1973. Gorokh. [Pea]. Kolos, L., 312 p. [Russian].
Maltzev, A.I. 1909. Otchet Buro po Prikladnoi Botanike za 1908 god. [Report of the Bureau of Applied Botany for 1908.] Bull. of the Bureau of Applied Botany 2(3):571-573. [Russian].
Maltzev, A.I. 1910. Otchet Buro po Prikladnoi Botanike za 1909 god. [Report of the Bureau of Applied Botany for 1909.] Bull. of the Bureau of Applied Botany 3(3/4):178-182. [Russian].
Maltzev, A.I. 1914. Iz nabludenii nad razvitiem dikorastushikh i sornykh ovsov. [On the development of wild and weedy oats]. Bull. of the Bureau of Applied Botany 7(12):786-791. [Russian].
Maltzev, A.I. 1916a. Otchet Buro po Prikladnoi Botanike za 1915 god. [Report of the Bureau of Applied Botany for 1915.] Bull. of the Bureau of Applied Botany 9(5):245-252. [Russian].
Maltzev, A.I. 1916b. Zasorennost posevov v Novgorodskoi gubernii. [Weeds in the gouv. of Novgorod]. Bull. of the Bureau of Applied Botany 9(4):137-174. [Russian].
Maltzev, A.I. 1930. Ovsyugi i ovsy. Sectio Euavena Griseb. [Wild and Cultivated Oats. Sectio Euavena Griseb.]. Works of Applied Botany and Plant Breeding. Supplement no. 38, [Russian, English].
Medvedev, Zh. A. 1969. The rise and fall of T.D. Lysenko. Columbia University Press, New York, London, 287 p. [English].
Merezhko, A.F. 1984. Sistema geneticheskogo izucheniya iskhodnogo materiala dlya selektzii rastenii. [The system of the genetic evaluation of the initial materials for plant breeding]. Leningrad, 70 p. [Russian].
Merezhko, A.F. 1994a. Vavilov Institute's collection of wheats and Aegilopses provides global food security. Diversity 10, no. 3. [English].
Merezhko, A.F. 1994b. Genetic analisis of quantitative traits for solution of plant breeding problems. Russian J. Genetics 30(10):1317- 1326. [English]
Merezhko, A.F. 1994c. Problema donorov v selektzii rastenii. [The problem of donors in plant breeding]. St. Petersburg, 126 p. [Russian].
Mesherov, E.T. and K.A. Kobylyanskaya. 1981. Rastitelnye resursy Sovetskogo Coyuza (ekspeditzii VIR v 1971-1980 godakh). [Plant resources of Soviet Union (VIR's expeditions in 1971-1980)]. N 322. L. 126 p. [Russian].
Mesherov, E.T. and K.A. Kobylyanskaya. 1982. Informatzia o zarubezhnykh ekspediztiyakh i komandirovkakh VIR v 1971-1980 gg. [Information on foreign expeditions and business trips of VIR in 1981-1985]. N 356. L. 40 p. [Russian].
Methodological guidance directory for studing VIR's collections. 1973. Leningrad, 148 p. [Russian].
Mitrofanova, O.P. 1994. Creation of a genetic collection of common wheat in Russia: a basis for further development of particular genetics and breeding. Russian J. Genetics 30(10):1306-1317. [English].
Muller, G. 1966. A letter to M. Popovsky of 16 July 1966. Personal Yu. N. Vavilov's archive. [English].
Podyma, W., S. Goral and H. Czembor. 1993. Conservation of genetic resources in Poland. Report of a Working Group on Management of Genetic Resources, Prague. [English].
Popovsky, M.A. 1984. The Vavilov Affair.The Shoe String Press, Inc., Hamden, Connecticut, USA. [English]

Popovsky, M.A. 1990. Delo akademika Vavilova. [The case of Academician Vavilov]. "Kniga", M., 303 p. [Russian].
Regel, R.E. 1906. Les orges cultivées de l'Empire Russe. Publication of the Department of Land Cultivation, St. Petersburg. [French].
Regel, R.E. 1915. Organizatzia i deyatelnost Buro po prikladnoi Botanike za pervoe dvadtzatiletie ego sushestvovaniya. [Organization and activity of the Bureau of Applied Botany for the first twenty years]. Bull. of the Bureau of Applied Botany 8(4/5):327-767. [Russian].
Regel, R.E. 1917a. Knyaz Boris Borisovich Golitzin [Prince B.B. Golitzin]. Bull. of the Bureau of Applied Botany 10(1):1-9. [Russian].
Regel, R.E. 1917b. K voprosu o vidoobrazovanii. [On the species formation]. Bull. of the Bureau of Applied Botany 10(1):157-181. [Russian].
Regel, R.E. 1922. Khleba Rossii. [Cereals of Russia]. Works on Applied Botany and Plant Breeding, Supplement no. 22, 55 p. [Russian].
Research Institute of Crop Production. 1992. Annual Report. Prague, Czech Republic. [English].
Reznik, S.E. 1968. Nikolai Vavilov. Moscow, 333 p. [Russian].
Rigin, B.V. 1994. The begining and development of genetic research in Vavilov All-Russian Institute of Plant Industry. Russian J. Genetics 30(10):1283-1293. [English].
Rukovodstvo po aprobatzii kulturnykh rastenii. 1938. Kormovye kultury. [Cultivated plant trials. Forages]. Vol. 4, M.-L., 222 p. [Russian].
Rukovodstvo po aprobatzii kulturnykh rastenii. 1938. Maslichnye kultury. [Cultivated plant trials. Oil crops]. Vol. 3, M.-L., 206 p. [Russian].
Rukovodstvo po aprobatzii kulturnykh rastenii. 1938. Zernovye i bobovye kultury. [Cultivated plant trials. Cereals and legumes]. Vol. 2, M.-L., 300 p. [Russian].
Rukovodstvo po aprobatzii kulturnykh rastenii. 1938. Zernovye kultury. [Cultivated plant trials. Cereals]. Vol. 1, M.-L., 510 p. [Russian].
Rukovodstvo po aprobatzii kulturnykh rastenii. 1939. Ovoshnye i kormovye korneplody. [Cultivated plant trials. Vegetable and root crops]. Vol. 5, M.-L., 563 p. [Russian].
Satsiperov, F. 1916. Opyt skreshivaniya dvukh form podsolnichnika *Helianthus annus* L. X *H. agrophyllus* .A.Gray. [The hybridization of *Helianthus annus* L. X *H. agrophyllus* A.Gray.] Bull. of the Bureau of Applied Botany 9(5):207-244. [Russian].
Sazonova, L.V., E.I. Gaevskaya and T.K. Lassan. 1994. VIR: proshloe I nastoyashee. [VIR: past and present]. Vestnik of the Russian Academy Agricultural Science 3:12-17. [Russian].
Sazonova, L.V. and E.A. Vlasova. 1990. Korneplodnye rastenia. [Root Crops]. Leningrad, 296 p. [Russian].
Shcherbakov, Yu.N. and V.A. Chikova. 1971. Ekspeditzii instituta po SSSR. [The expeditions of the Institute within the USSR]. Works of Applied Botany, Genetics and Plant Breeding 45(2):299-320. [Russian].
Shcherbakov, Yu.N. and V.A.Chikova. 1970. Zarubezhnye ekspediztii VIRa po sboru rastitelnykh resursov. [The foreign expeditions of VIR]. Works of Applied Botany, Genetics and Plant Breeding 42(2):316- 320. [Russian].
Sinskaya, E.N. 1969. Istoricheskaya geografia kul'turnoi flory (na zare zemledeliya). [Historical geography of cultivated plants (at the dawn of agriculture]. "Kolos", L., 480 p. [Russian].
Sinskaya, E.N. 1991. Vospominaniya o N.I.Vavilove [Memories about N.I.Vavilov]. Naukova Dumka, Kiev. 209 p. [Russian].
Soifer, V.N. 1993. Vlast i nauka. Istoriya razgroma genetiki v SSSR. [The power and the science. [The history of the ruin of genetics in the USSR]. Moscow, 706 p. [Russian].
Soratniki Nikolaya Ivanovicha Vavilova. Issledovateli genofonda rastenii. [Nikolai Ivanovich Vavilov's associates. Researchers of the plant genepool]. 1994. St. Petersburg, 607 p. [Russian].
Stehno, Z. and V. Skaloud. 1991. Czecho-Slovak gene bank of cultivated plants. Plant genetic

Resources. Czechoslovak board on plant genetic resources. Annual report, p. 53-59. [English, Czechoslovakian].
Teoreticheskie osnovy selektzii rastenii. 1935. Chastnaya selektziya zernobykh i kormovykh kultur. [Theoretical bases of plant breeding. Special part. Grain and forage crops]. Vol. II, M.-L., 712 p. [Russian].
Teoreticheskie osnovy selektzii rastenii. 1935. Obshaya selektziya rastenii. [Theoretical bases of plant breeding. General principles of plant breeding]. Vol. I, M.-L., 1043 p. [Russian].
Teoreticheskie osnovy selektzii rastenii. 1937. Chastnaya selektziya kartofelya, ovoshnykh, bakhchevykh, plodovo-yagodnykh i tekhnicheskikh kultur. [Theoretical bases of plant breeding. Special part. Potatoes, vegetables, *Cucurbitaceae*, fruits, small fruits and industrial crops]. Vol. III, M.-L., 862 p. [Russian].
Trofimovskaya, A.Ya. 1972. Yachmen. (Evolutziya, klassifikatziya, selectziya) [Barley. Evolution, taxonomy, breeding]. Kolos, L., 296 p. [Russian].
Udachin, R.A. and I.Sh. Shakhmedov. 1984. Pshenitzy v Srednei Azii. [Wheats in Middle Asia]. Tashkent, 135 p. [Russian]
Vavilov, N.I. 1915. Immunity to fungus diseases as a physiological test in genetics and systematics, exemplified in cereals. J. Genetics 4:49-65. [English].
Vavilov, N.I. 1917. O proiskhozhdenii kulturnoi rzhi [On the origin of the cultivated rye]. Bull. of the Bureau of Applied Botany 10(7-10):561-590. [Russian].
Vavilov, N.I. 1918. Immunitetrastenii k infektzionnym zabolevaniyam. [Plant immunity to infectious diseases]. Works of the Petrovsky Agricultural Academy 1-4, 240 p. [Russian].
Vavilov, N.I. 1920. Zakon gomologicheskikh ryadov. [The law of homologous series in variation]. Proceedings of the III All-Russian plant breeding conference. Saratov, 16 p. [Russian].
Vavilov, N.I. 1922a. The law of homologous series in variation. J. Genetics XII:47-89. [English].
Vavilov, N.I. 1922b. Letters to W. Bateson. Archives, John Innes Center. [English].
Vavilov, N.I. 1922c. Polevye kultury Yugo-Vostoka. [Field crops of the South-East]. Works of Applied Botany and Plant Breeding. Supplement no. 23, 228 p. [Russian].
Vavilov, N.I. 1923. K poznaniyu myagkikh pshenitz.(Sistematiko-geograficheskii ocherk). [On the congnition of bread wheat [Taxonomic and geographical essay]. Works of Applied Botany and Plant Breeding 13(1):149-257. [Russian].
Vavilov, N.I. 1924. O vostochnykh tzentrakh proiskhozhdeniya kulturnykh rastenii. [On the Eastern centres of origin of cultivated plants]. Novyi Vostok 6:291-305. [Russian].
Vavilov, N.I. 1926a. Tzentry proiskhozhdeniya kulturnykh rastenii. [The centers of origin of cultivated plants]. Works of Applied Botany and Plant Breeding 16(2), 248 p. [Russian, English].
Vavilov, N.I. 1926b. William Bateson. 1861-1926. Works of Applied Botany and Plant Breeding 15(5):499-511. [Russian].
Vavilov, N.I. 1927a. Essais geografiques sur l'etude de la variabilite des plantes cultivées en l'URSS (Russie). Int. Rev. Agric. 18(11):630-664.
Vavilov, N.I. 1927b. Geograficheskie zakonomernosti v raspredelenii genov kulturnykh rastenii. [Geographical regularities in relation to the distribution of the genes of cultivated plants]. Works of Applied Botany, Genetic and Plant Breeding 17(3):411-428. [Russian].
Vavilov, N.I. 1928a. Les centres mondiaux des genes du ble. Actes de la 1-ere Conference internationale du ble. Roma, 1927:368-376.
Vavilov, N.I. 1928b. Geographische Genzentren unserer Kulturpflanzen. Verhandlungen des V Internationalen Kongresses fur Vererbungswissenschaft. Berlin, 1927, Bd. 1., S.342-369.
Vavilov, N.I. 1930. Science and technique under conditions of a socialist reconstraction of agriculture. International conference of agriculture economists, Ithaca, New York, p. 1-14. [English].
Vavilov, N.I. 1931a. The Linnean species as a system. Report of proceedings of V International

Botanical Congress, Cambridge, 1930, p. 213-216. [English].
Vavilov, N.I. 1931b. Wild progenitors of the fruit trees of Turkistan and the Caucasus and the problen of the origin of fruit trees. Proceeding of IX International Horticultural Congress, London, 1930, p. 217-286. [English].
Vavilov, N.I. 1931c. The problem concerning the origin of agriculture in the light of recent research. International congress of the history of science and technology, London, p. 95-106. [English].
Vavilov, N.I. 1931d. Rol Tzentralnoi Azii v proiskhozhdenii kulturnykh rastenii. [The role of Central Asia in the origin of cultivated plants]. Works of Applied Botany, Genetic and Plant Breeding 26(3):3-44. [Russian, English].
Vavilov, N.I. 1931e. Meksika i Tzentralnaya Amerika kak osnovnoi tzentr proiskhozhdeniya kulturnykh rastenii Novogo Sveta. [Mexico and Central America as a basic centre of origin of cultivared plants in the New World]. Works of Applied Botany, Genetic and Plant Breeding 26(3):135-199. [Russian, English].
Vavilov, N.I. 1932. The process of evolution in cultivated plants. Proc. 6th International Congress of Genetics. Ithaca, NY, Vol. 1:331-342. [English].
Vavilov, N.I. 1933a. L'agriculture et la science agronomique en URSS. Rev. int. bot. 13(140):241-251.
Vavilov, N.I. 1933b. Das Problem der Entstehung der Kulturpflanzen. Nova Acta Leopoldina carol., N.F., Bd. 1, H. 2-3, S.332-337.
Vavilov, N.I. 1935. A letter to O. Frankel of August 1935. Personal Yu.N.Vavilov's archive. [English].
Vavilov, N.I. 1939. Genetics in the USSR. Chronica Botanica 5(1):14-15. [English].
Vavilov, N.I. 1940. The new systematics of cultivated plants. Pp. 549-566 *in* The New Systematics (J. Huxley, ed.). Claredon Press, Oxford. [English].
Vavilov, N.I. 1941. Entering a new epoch. Chronica Botanica 6(19/20):433-437. [English].
Vavilov, N.I. 1951. The origin, variation, immunity and breeding of cultivated plants. Transl. by K.S. Chester. Ronald Press, NY, 366 p. [English].
Vavilov, N.I. 1957a. Mirovye resursy sortov khlebnykh zlakov, zernovykh bobovykh, l'na i ikh ispolzovanie v selektzii. Opyt agroekologicheskogo obozreniya vazhneishikh polevykh kultur. [World resources of cereals, leguminous seed crops and flax, and their utilization in plant breeding. Agro-ecological survey of the principal field crops]. USSR Academy of Science Press, M.-L., 462 p. [Russian, English].
Vavilov, N.I. 1957b. Gornoe zemledelie Severnogo Kavkaza i perspektivy ego razvitiya. [Mountainos agriculture of the North Caucasus and prospects of its development]. Herald of the Russian Academy of Science. Biology 5:590-600. [Russian].
Vavilov, N.I. 1959. Izbrannye trudy. Zemledelcheskii Afganistan. [Selected works. (Agricultural Afghanistan]. Vol. 1. USSR Academy of Science Press, M.-L., 416 p. [Russian].
Vavilov, N.I. 1960. Izbrannye trudy. Problemy selektzii, rol Evrazii i Novogo Sveta v proiskhozhdenii kulturnykh rastenii. [Selected works. The problems of breeding, the role of EurAsia and New World in origin of cultivated plants]. Vol. 2. USSR Academy of Science Press, M.-L., 520 p. [Russian].
Vavilov, N.I. 1962. Izbrannye trudy. Problemy geografii, filogenii i selektzii pshenitzy i pzhi. Rastitelnye resursy i voprosy sistematiki kulturnykh rastenii. [Selected works. The problems of geography, taxonomy and breeding of wheat and rye. Plant resources and problems of the taxonomy of cultivated plants]. Vol. 3. USSR Academy of Science Press, M.-L., 532 p. [Russian].
Vavilov, N.I. 1964. Mirovye resursy sortov khlebnykh zlakov, zernovykh bobovykh, l'na i ikh ispolzovanie v selektzii. Pshenitza. [World resources of cereals, leguminous seed crops and flax, and their utilization in plant breeding. Wheats]. Nauka Press, M.-L., 122 p. [Russian].
Vavilov, N.I. 1965. Izbrannye trudy. Problemy proiskhozhdeniya, geografii, genetiki, selektzii

rastenii, rastenievodstva i agronomii. [Selected works. The problems of origin, geography, genetics, plant breeding, plant industry and agronomy]. Vol. 5. USSR Academy of Science Press, M.-L., 788 p. [Russian].
Vavilov, N.I. 1980. Nauchnoe nasledstvo [Scientific inheritance]. Vol. 5. Vavilov's letters 1911-1928, Nauka Press, Moscow, 428 p. [Russian].
Vavilov, N.I. 1987a. Nauchnoe nasledstvo [Scientific inheritance]. Vol. 10. Vavilov's letters 1929-1940, Nauka Press, Moscow, 494 p. [Russian].
Vavilov, N.I. 1987b. Pyat kontinentov [Five continents]. Nauka Press, Moscow. 171 p. [Russian].
Vavilov, N.I. 1987c. Ocherki, materialy, dokumenty. [Issues, materials, documents]. Nauka Press, Moscow, 487 p. [Russian].
Vavilov, N.I. 1990. "Zhizn korotka, nado speshit". ["Life is short, so we must hurry"]. Moscow, 704 p. [Russian].
Vavilov, N.I. 1992. Origin and Geography of Cultivated Plants. Transl. by Doris Love. Cambridge University Press, 498 p. [English].
Vavilov, N.I. 1994. Nauchnoe nasledie v pismakh. Mezhdunarodnaya perepiska. [Scientific heritage in letters. International correspondence]. 1921-1927, Vol. I. Nauka Press, Moscow, 556 p. [Russian].
Vavilov, N.I. 1997. Five continents. Transl. by Doris Love. IPGRI/VIR, Italy, Rome, 198 p. [English].
Vitkovsky, V.L. 1984. Morfogenez plodovykh rastenii. [Morphogenesis of fruit plants]. Kolos, L., 206 p. [Russian].
Vitkovsky, V.L., S.V.Kuznetsov. 1990. The N.I.Vavilov All-Union Research Institute of Plant Industry. Diversity 6(1):15. [English].
Vosazhdennom Leningrade. [In the siege Leningrad]. 1969, Lenizdat, Leningrad, 147 p. [Russian].
Vulf, E.V. 1944. Istoricheskaya geografiya rastenii. Istoriya flor zemnogo shara. [Historical geography of the plants. The history of the world's floras]. M.-L., 545 p. [Russian]
Westover, H.L. 1929. Report of the expedition to the USSR. National Archives of the USA. [English].
Westover, H.L. 1934. Letters from the USSR. National Archives of the USA. [English].
Zaitzev, V.A. 1990. Long-term seed storage in the USSR: modern genebank at the Kuban seed testing station. Diversity 6(2):14. [English].
Zeven, A.C. and P.M. Zhukovsky. 1975. Dictionary of cultivated plants and their centres of Diversity. Centre for Agricultural Publishing and Documentation, Pudoc, The Netherlands. [English]
Zhukovsky, P.M. 1964. Kulturnye rasteniya i ikh sorodichi. [Cultivated plants and their relatives]. Kolos, L., 791 p. [Russian].
Zhukovsky, P.M. 1967. Image of N.I. Vavilov. *In* N.I. Vavilov. Izbrannye trudy [Selected works]. Vol. 2. Nauka Press, Leningrad, p. 439-453. [Russian].
Zhukovsky, P.M. 1970. Mirovoi genofond rastenii dlya selektzii (megatzentry i mikrotzentry). [Worldwide genebank of plants for plant breeding (mega-centers and micro-centers]. Nauka Press, L., 88 p. [Russian]
Zoteeva, N.M. and T.A. Komarova. 1996. Informatzia ob ekspediztiyakh VIR, provedennykh na territorii Rossii, stran blizhnego i dalnego zarubezh'ya v 1986-1994 gg. [Information on expeditions VIR within Russia, CIS and foreign countries in 1986-1994]. N 682. St. Peterburg. 66 p. [Russian].
Zukovsky, P.M. 1934. Zemledelcheskaya Turtziya. (Aziatskaya chast – Anatolia). [Agricultural Anatolia]. M.-L., 907 p. [Russian, English].